Laboratory Manual

Fluid Power

Hydraulics and Pneumatics

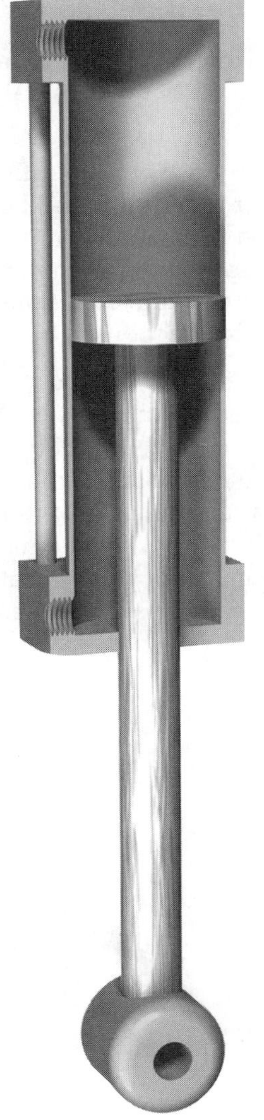

by

James R. Daines
Professor Emeritus
University of Wisconsin—Stout
Menomonie, WI

Publisher
The Goodheart-Willcox Company, Inc.
Tinley Park, Illinois
www.g-w.com

Manufactured in the United States of America.

ISBN 978-1-60525-082-3

1 2 3 4 5 6 7 8 9 – 09 – 14 13 12 11 10 09

Introduction

This laboratory manual supports the textbook *Fluid Power: Hydraulics and Pneumatics.* The activities contained within range from simple tasks related to instruction about basic fluid concepts to advanced principles concerning component parts and their operation in hydraulic or pneumatic circuits. The activities have been selected and designed to help individuals seeking general information, suitable information to become fluid power technicians, or an engineering degree.

The manual includes a variety of hands-on activities:

- Guides for the observation of various fluid power applications.
- Study of manufacturer component sheets.
- Component testing.
- Circuit assembly and operation.

The activities have not been developed using any specific fluid power training unit. Many of the activities can be adapted to commercial training units.

Each laboratory activity includes an introduction that describes the assignment and provides detailed specifications, including a schematic diagram when appropriate. Required components are identified with notes indicating their use in the activity. Detailed, step-by-step procedures are provided along with charts to record resulting data and observations. The activities conclude with activity analysis questions. Some analysis questions may require you to synthesize information from several locations (laboratory tests, textbook, Internet, and manufacturer resources) to obtain a correct answer.

—Jim Daines

Warning: Pressurized liquids and gases need to be carefully handled to prevent injuries. Always handle hydraulic and pneumatic systems with respect. Be certain the system prime mover is turned off and system line pressure is at zero before attempting to assemble or disassemble test circuits suggested for the activities.

Table of Contents

	Textbook page	Lab Manual page
Chapter 1 **Introduction to Fluid Power**	**15**	**7**
Chapter 2 **Fluid Power Systems**	**33**	**19**
Chapter 3 **Basic Physical Principles**	**51**	**37**
Chapter 4 **Fluid Power Standards and Symbols**	**83**	**59**
Chapter 5 **Safety and Health**	**113**	**77**
Chapter 6 **Hydraulic Fluid**	**139**	**91**
Chapter 7 **Source of Hydraulic Power**	**167**	**99**
Chapter 8 **Fluid Storage and Distribution**	**209**	**117**
Chapter 9 **Actuators**	**233**	**133**
Chapter 10 **Controlling the System**	**273**	**155**
Chapter 11 **Accumulators**	**331**	**187**
Chapter 12 **Conditioning System Fluid**	**349**	**209**
Chapter 13 **Applying Hydraulic Power**	**369**	**223**
Chapter 14 **Compressed Air**	**393**	**261**

	Textbook page	Lab Manual page
Chapter 15 **Source of Pneumatic Power**	**405**	**287**
Chapter 16 **Conditioning and Distribution of Compressed Air**	**425**	**309**
Chapter 17 **Work Performers of Pneumatic Systems**	**451**	**323**
Chapter 18 **Controlling a Pneumatic System**	**475**	**345**
Chapter 19 **Applying Pneumatic Power**	**497**	**367**

Chapter 1

Introduction to Fluid Power

The Fluid Power Field

The activities in this lab exercise are designed to show applications of fluid power as discussed in the text. The activities are based on observations made in a variety of everyday living and/or working situations. Each activity is designed to direct the observer to often-overlooked aspects of fluid power that are found in our culture. Be certain to obtain permission to make observations in laboratories and any businesses you may visit as you conduct your work.

> **Note:** Check with your instructor to be certain that all the policies of the school are followed, especially if your activities involve a group other than the school.

Key Terms

The following names and terms are used in this chapter. As you read the text, record the meaning and importance of each. Additionally, you may use other sources, such as manufacturer literature, an encyclopedia, or the Internet, to obtain more information.

accumulator ______

actuator ______

Archimedes ______

Armstrong, Lord William ______

Bernoulli, Daniel ______

Boyle, Robert ______

Boyle's law ______

Bramah, Joseph ______

Charles, Jacques ______

Charles' law ______

compact hydraulic unit ______

cup seal ______

da Vinci, Leonardo ______

fluid compressibility ______

fluid power ______

general gas laws ______

Hero ______________________________

____

____

hydraulic ______________________________

____

____

hydraulic accumulator ______________________________

____

____

hydraulic intensifier ______________________________

____

____

Industrial Revolution ______________________________

____

____

laminar ______________________________

____

____

Maudslay, Henry ______________________________

____

____

Pascal, Blaise ______________________________

____

____

pneumatic ______________________________

____

____

prime mover ______________________________

____

____

pump ______________________________

____

____

Reynolds number ______________________________

____

____

Reynolds, Osborne ______________________________

____

____

sails ______________________________

____

____

scientific method ______________________________

____

____

steam engine ______________________________

____

____

Torricelli, Evangelista ______________________________

____

____

turbulent ______________________________

____

____

von Guericke, Otto ______________________________

____

____

water screw ______________________________

____

____

waterwheel ______________________________

____

____

Watt, James ______________________________

____

____

windmill ______________________________

____

____

Chapter 1 Quiz

Name ______________________ **Date** ____________ **Score** ________

Write the best answer to each of the following questions in the blanks provided.

1. The two broad classifications of fluid power systems are ______ and ______.

2. Name the five areas that contribute to the success of the fluid power industry.

3. List the basic system characteristics where differences may be found between hydraulic and pneumatic systems.

4. The ______ fluid power system is usually selected when an application requires a high operating pressure.

5. The most accurate component speed control is provided by the ______ fluid power system.

6. Both hydraulic and pneumatic systems provide an easy means of multiplying and controlling ______ and ______.

7. Archimedes is credited with the discovery of the scientific principle of ______. He also invented the ______, which has been used for centuries to lift and move water.

8. Three items essential to human existence and comfort that have been closely related throughout history to the development of fluid power are:

9. Waterwheels were first developed during the ______ century BC.

10. Average-size windmills typical of the Middle Ages developed only ______ to ______ horsepower.

11. The changes that occurred during the Industrial Revolution were not only dependent on the development of mechanical devices, but also involved:

12. Many of the concepts involved in current fluid power systems were first used during the Industrial Revolution to remove water from ______.

____

13. The development of the ______ in the late 1700s allowed cylinders and other devices to be constructed that did not leak and could be continuously used.

____

14. Centralized pumping stations were commonly used in Britain in the late 1800s and early 1900s to distribute pressurized water up to ______ miles.

____

15. Compact, self-contained hydraulic power units were first introduced in the ______.

____

Activity 1-1

Identification of Equipment and Products Associated with Fluid Power Products

Name ________________________________ **Date** ______________

This activity is designed to show how fluid power is an often-overlooked aspect of everyday life. To emphasize this, you will examine brochures often included as advertising materials in local newspapers. If you are not familiar with these materials, check with your instructor for a recommendation.

Activity Specifications

Select a home, construction, farm product, or merchandise advertising flyer or brochure from the newspaper in your area. Identify the products and replacement parts listed in the publication that are in some way associated with fluid power. Complete the questions below based on your research.

1. Source of the advertising brochure: ______________________________
   Number of pages in the brochure: ______________________________
   Estimated total number of products listed in the brochure: ______________
2. Estimate the number of products associated with the following general groups:
   Group 1, construction tools and equipment: ______________________
   Group 2, lawn and farm tools and equipment: ______________________
   Group 3, household and kitchen appliances: ______________________
   Group 4, automotive and truck replacement parts and service products: ______________
3. Examine the products from each of the above categories, considering the following points. List the product names.
   A. Is fluid power (hydraulic or pneumatic) used directly in the operation of the product?
   ____
   ____
   ____
   B. Are basic fluid mechanic principles, such as air and fluid flow and pressure, used in the operation of the product or component, even though it might not be an easily visible part of the product's operation?
   ____
   ____
   ____
   C. Are any of the advantages of fluid power used to promote the usefulness and desirability of the product?
   ____
   ____
   ____

4. How many of the products or components identified in the four categories are associated with fluid power?

   Group 1: ____________________

   Group 2: ____________________

   Group 3: ____________________

   Group 4: ____________________

5. Select one product from each group and describe how fluid power elements are involved in product operation or the promotion of the product.

   Group 1: ____________________

   ____

   ____

   ____

   Group 2: ____________________

   ____

   ____

   ____

   Group 3: ____________________

   ____

   ____

   ____

   Group 4: ____________________

   ____

   ____

   ____

6. From the analysis of the products listed in the brochure, describe the influence of fluid power on consumer products:

   ____

   ____

   ____

   ____

Activity 1-2

Identification of Tools and Equipment Used at a Selected Construction Site Employing Fluid Power

Name ______________________________ **Date** ______________

This activity is designed to show how fluid power is used in the construction industry, including the building of homes, commercial properties, streets, and highways. Work with your instructor to select a site to visit. When requesting permission to visit a building site, it should be indicated that you will observe all regulations required by the company. Also, be certain to follow all of the policies of your school regarding outside activities.

Activity Specifications

Select a building, street, or highway construction site. Visit the site to observe fluid power equipment in use and the activities of the workers using the equipment. Identify all equipment and products that are in some way associated with fluid power. Complete the questions below based on the site visit.

1. Location of site:

   ____

2. Describe the type of construction underway at the site (single-family home, office building, street paving, factory, irrigation system, etc.):

   ____

3. List the major pieces of equipment observed at the site:

   ____

   ____

4. What percentage of the equipment observed at the site appears to use fluid power for at least part of its operation? ______ List the names of equipment using fluid power:

   ____

   ____

   ____

5. Select one piece of equipment identified in item 4 and describe how fluid power appears to be used in its operation.

   ____

   ____

   ____

6. List the portable power tools observed at the site. Attempt to identify the power source (pneumatic, hydraulic, electrical, or unknown) for these tools.

Tool	Power Source
______________	______________
______________	______________
______________	______________
______________	______________
______________	______________

7. For the pneumatic and hydraulic tools identified in item 6, which characteristics make fluid power desirable as a source of power?

8. Based on your observations, describe the influence fluid power has on the construction industry:

Activity 1-3

Observation of Fluid Power Applications Related to Residential Support Services

Name ______________________________________ Date ____________

This activity is designed to show how fluid power principles and equipment are used in residences. The residence may be a single-family home, townhouse, or multiunit apartment or condominium complex. Be certain to request permission to visit the site from the homeowners and, for larger complexes, the governing association. When requesting permission to visit the site, be certain to indicate that you will observe all regulations required by the owners. Also, be certain to follow all of the policies of your school regarding outside activities.

Activity Specifications

Select a residential facility ranging from a small single-family home to a multiunit housing complex. Visit the site to identify the fluid power concepts used in the building(s) and the equipment needed to maintain the facility. Identify all equipment and products that are in some way associated with fluid power. Use the activity questions below to be certain all items are considered as you observe the site. Complete the questions below based on the site visit.

1. Generally describe the residence you are observing for this activity (single-family home, duplex, apartment, etc.):
   ____

2. Describe the system(s) used to heat, ventilate, and cool the residence (forced air, hot water, steam, etc.):
   ____
   ____

3. Describe how the design and operation of the system in the residence uses component parts and basic concepts (flow and pressure) that are closely associated with fluid power:
   ____
   ____
   ____

4. Identify the source of water for the residence (public water system, private well, etc.):
   ____

5. Describe how fluid mechanic and fluid power concepts (flow and pressure) are involved in the operation of the water system:
   ____
   ____

6. List the major household appliances that are used in the residence:
   ____
   ____
   ____
   ____

7. Identify two appliances from item 6 that use fluid power concepts in their operation. Describe those concepts and how they are used in the operation of the appliance.
A. ______________________________
____
____
____
B. ______________________________
____
____
____

8. From your observations, describe the influence of fluid power concepts on the residence and the way the residents perform the everyday tasks:
____
____
____
____

Activity 1-4

Observation of Fluid Power Applications Related to the Support and Operation of a Business

Name ______________________ **Date** __________

This activity is designed to show how fluid power principles and equipment are used in the physical facilities of a typical business operation. The organization selected should be a small- or medium-size business in the retail/wholesale, service, manufacturing, or healthcare areas. Be certain to request permission to visit the site of the operation. When requesting permission to visit the site, indicate that you will observe all regulations required by the owner/operator/manager. Also, be certain to follow all of the policies of your school regarding outside activities.

Activity Specifications

Choose a field of special interest to you. Identify a local group/company functioning in this field. Contact the owner/manager to obtain permission to visit the facility to identify fluid power concepts and equipment used in their business operations. Visit the site with the specific objective of identifying how the fluid power field is affecting the business operation. During the visit, carefully examine all building facilities, equipment, and products for any connection to fluid power concepts. Complete the questions below based on the observations made during the site visit.

1. Give the name and location of the business or organization visited:
____

2. Briefly describe the purposes of the organization and the general activities performed at the site:
____
____
____

3. List general areas in which you observed the use of fluid power in the organization. Identify these areas in broad general terms, such as materials handling, communication, production, and so on.
____
____
____

4. Select two of the areas identified in item 3 and describe the type of fluid power applications observed.
A. ______________________
____
____
____
B. ______________________
____
____
____

5. Which basic characteristics made the use of fluid power desirable for the applications selected in item 4?

6. Describe two additional fluid power applications that would benefit the operation of the organization. Do not include those identified in item 5.

A.

B.

7. What benefits would the organization experience if your suggestions in item 6 are incorporated?

8. From your observations, describe the influence of fluid power concepts on the organization:

9. How would you rate the potential for continued and/or expanded use of fluid power in this organization?

10. How would you rate the potential of fluid power in the general field represented by the organization you observed for this activity? Why?

Chapter 2

Fluid Power Systems

The Basic System

The activities in this lab exercise are designed to show details of basic fluid power systems and circuits. The first activities involve analysis of systems in operation on hydraulic- and pneumatic-powered equipment. The remaining activities involve the assembly and simple testing of basic circuits. Each activity is designed to direct you to fundamental concepts of hydraulic and pneumatic components and circuits. Be certain to obtain permission to make observations in laboratories and any businesses you may visit as you conduct your work.

Key Terms

The following terms are used in this chapter. As you read the text, record the meaning and importance of each. Additionally, you may use other sources, such as manufacturer literature, an encyclopedia, or the Internet, to obtain more information.

actuators ____________________

adapters ____________________

component groups ____________________

compressor ____________________

conductors ____________________

control valves ____________________

cylinders ____________________

directional control valves ____________________

filters ____________________

fittings ____________________

flow control valves ____________________

fluid conductor ____________________

fluid maintenance devices ____________________

heat exchanger ____________________

hoses ____________________

lubricator ____________________

mechanical coupler ____________________

motors ____________________

pipes ____
____
____

power unit ____
____
____

pressure control valves ____
____
____

pressure regulator ____
____
____

prime mover ____
____
____

pump ____
____
____

receiver ____
____
____

reservoir ____
____
____

separator ____
____
____

system functions ____
____
____

tubes ____
____
____

Chapter 2 Quiz

Name ______________________ Date __________ Score ______

Write the best answer to each of the following questions in the blanks provided.

1. Component groups designed to perform specific ______ make up fluid power systems.

 ____

2. The ______ function of a fluid power system transforms prime mover energy into a form that a(n) ______ can use to perform work.

 ____

3. Name the three tasks associated with the fluid conditioning/fluid maintenance function of fluid power systems.

 ____

4. The power unit for basic fluid power systems consists of five components. List these for both hydraulic and pneumatic systems.

Hydraulic System	Pneumatic System
____________	____________
____________	____________
____________	____________
____________	____________
____________	____________

5. A(n) ______ is an actuator that converts energy stored in the system fluid into linear motion.

 ____

6. When the connection between fluid power components in a fluid system must be flexible, a(n) ______ is usually the best solution.

 ____

7. Name the three control tasks performed by pressure control valves in fluid power systems.

 ____

 ____

 ____

8. In a hydraulic system, fluid that is not needed to maintain desired operating pressure is returned to a(n) ______ section of the system through the pressure control valve.

 ____

9. The basic pressure control device that sets maximum workstation operating pressure in a pneumatic system is called a(n) ______.

 ____

10. Proper maintenance of system fluid is important for both the ______ and ______ of components in a fluid power system.

 ____

11. A(n) ______ valve is used in the design of fluid power system filters to route fluid around the filter element if it becomes clogged during system operation.

 ____

12. In a pneumatic system, a filter is often combined with a(n) ______ and called a(n) ______.

 ____

13. The ______ adds a fine mist of oil to a pneumatic system to assure lubrication of system components.

____

14. In a hydraulic fluid power system, oil is returned to the ______ after it has completed its work at the actuator.

____

15. In a basic hydraulic system, actuator extension speed is controlled by placing a(n) ______ in the line between the directional control valve and the actuator.

____

Activity 2-1

Hydraulic Fluid Power System Observation and Analysis

Name ______________________________ Date ______________

This activity is designed to show the structure and operation of a hydraulic circuit on an operating machine. Ideally, the machine should include hydraulic power as a significant part of its operation. Your instructor should suggest suitable equipment and locations where it may be found.

Note: Check with your instructor to be certain that all the policies of the school are followed, especially if your activities involve a group other than the school.

Activity Specifications

Study the list of questions below to become familiar with the factors that will be observed and analyzed in the hydraulic system. Based on the factors that should be observed, select an appropriate piece of equipment to be studied. Identify a machine that appears to contain as many of these factors as possible. Obtain permission to observe the machine for an extended operating period. Complete the activity questions based on your observations.

1. In general terms, describe the machine and what it does.

2. Where is the machine located and who is the owner?

3. Estimate the percentage of machine functions operated by the hydraulic system. How important are these functions to the overall operation of the machine?

4. What type of prime mover is used to operate the hydraulic system of the machine? Identify the horsepower rating of the prime mover.

5. Describe the construction of the reservoir of the system power unit. Give the length, height, and width of the tank.

6. Locate the pump and identify its flow rate per minute. Describe how the pump is connected to the prime mover.

7. What fluid is used in the system?

8. Identify the filters used in the system. Describe the location of each filter.

9. Identify the maximum operating pressure of the system. What method appears to be used to adjust this pressure?

10. What types of actuators are used in the system? How many of each type are used and what is the general function of each?

11. What serves as the control valve to start and stop or reverse the actuators? Describe the appearance of the valve and how the machine operator manipulates it to obtain the desired machine operation.

12. What method appears to be used in the system to control actuator speed?

13. What types of fluid conductors are used in the system? What has been used to provide flexible connections between fluid power components and moveable machine parts?

14. Identify the methods used in the hydraulic system to maintain correct operating temperature. If no specific components can be identified, describe in general terms how temperature control may be accomplished.

15. Identify all of the component parts of the hydraulic system on the machine. Develop a pictorial diagram of the system. Be sure to included all conductors as well as the pump, valves, and actuators. Label each component with its correct name and function.

Activity 2-2

Pneumatic Fluid Power System Observation and Analysis

Name ______________________________ **Date** ______________

This activity is designed to show the structure and operation of a pneumatic circuit on an operating machine. Ideally, the machine should include pneumatic power as a significant part of its operation, but is not excessively complex. Your instructor should suggest suitable equipment and locations where it may be found.

> **Note:** Check with your instructor to be certain that all the policies of the school are followed, especially if your activities involve a group other than the school.

Activity Specifications

Study the list of questions below to become familiar with the factors that will be observed and analyzed in the pneumatic system. Based on the factors that should be observed, select an appropriate piece of equipment to be studied. Identify a machine that appears to contain as many of these factors as possible. Obtain permission to observe the machine for an extended operating period. If the machine obtains compressed air from a general air distribution system, rather than a compressor located on the machine, obtain permission to visit the central compressor unit. Complete the activity questions based on your observations.

1. In general terms, describe the machine and what it does.

2. Where is the machine located and who is the owner?

3. Estimate the percentage of machine functions operated by the pneumatic system. How important are these functions to the overall operation of the system?

4. What is the source of compressed air for the operation of the machine? If a central compressor unit is involved, where is the compressor located? Estimate the number of workstations supplied by the central air compressor.

5. Describe the prime mover used to operate the compressor unit. Identify the horsepower rating of the prime mover. How is the prime mover connected to the compressor?

6. Describe the size, shape, and construction of the receiver used with the compressor unit. Provide the dimensions of the diameter and length or height of the tank.

7. What is the maximum pressure of the compressed air in the receiver? Determine the operating range of air pressure in the receiver. What is the pressure when the compressor stops supplying air? What is the pressure when the compressor restarts and again supplies air? How is this operating range set for the compressor unit?

8. Identify the filters used in the system. Describe the location of each filter. Why do you think filters are used at these points in the pneumatic system?

9. Locate the pressure regulator at the system workstation. What is the pressure setting of the regulator? How does this working pressure compare to the pressure at the receiver? Why do you think this difference exists?

10. What types of actuators are used in the system? How many of each type are used and what is the general function of each?

11. What serves as the control valve to start and stop or reverse the actuators? Describe the appearance of the valve and how it is manipulated by the machine operator to obtain the desired machine operation.

12. What method appears to be used to control actuator speed in the system?

13. What types of fluid conductors are used in the system? What has been used to provide flexible connections between fluid power components and moveable machine parts?

14. Identify the methods used in the system to allow the removal of water from system lines. Describe the location of each device in the system.

15. Identify the components of the pneumatic system, including those that are part of the machine and the compressor unit. Be sure to include the air distribution system if the system has a central compressor. Develop a pictorial diagram of the system. Label each component with its correct name and function. Develop a list of the system functions.

Activity 2-3

Basic Hydraulic System Assembly and Operation

Name ______________________________ **Date** ______________

In this activity, you will work directly with actual hydraulic system components, including a hydraulic power unit, control valves, actuators, and the conductors used to transmit oil between components. The activity familiarizes you with system construction and operation, including the control of system pressure and actuator speed and direction.

Activity Specifications

Construct the basic hydraulic system shown in the following diagram. Operate the system and record observations of the simple tests described in this activity.

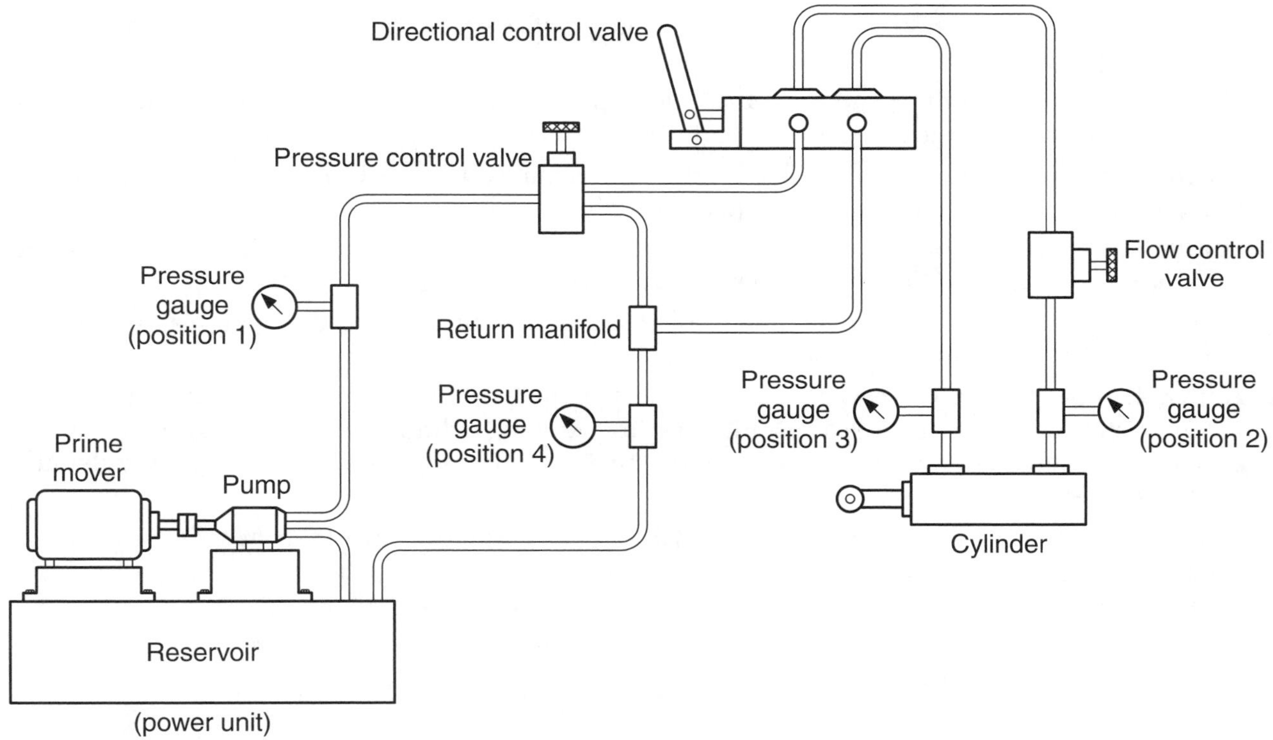

Required Components

Access to the following components is needed to complete this activity.

- Hydraulic power unit.
- Pressure control valve (relief valve).
- Directional control valve (four way, three position, closed-center spool, manually operated).
- Flow control valve (with return check valve).
- Cylinder (double acting).
- Pressure gauges (four).
- Manifolds (adequate number to allow supply, return, and gauge connections).
- Hoses (adequate supply to allow assembly of system).

Procedures

Study the following list of procedures to become familiar with the steps needed to complete the activity. Then, complete the procedure and record the observations indicated in the next section.

1. Select the components needed to assemble the system.
2. Assemble the system and have your instructor check the setup before proceeding.

> **Caution:** Set the pressure control valve at the *lowest* pressure setting and shift the directional control into the *center* position before starting the power unit.

3. Start the power unit and adjust the pressure control valve until the pressure gauge at position 1 reads 200 psi.
4. Shift the directional control valve lever to extend and then retract the cylinder to assure the system is operational.
5. Adjust the flow control valve until cylinder extension takes ten seconds. Time the retraction time of the cylinder. Record the time in Section A of the Data and Observation Records section.
6. Read and record the pressure at each of the gauge positions with the directional control valve in the center position.
7. Shift the directional control valve lever to extend the cylinder rod. Read and record the pressure at each of the gauge positions while the cylinder rod is extending.
8. When the cylinder is fully extended, hold the directional control valve lever in the shifted position and record the pressure at each gauge position.
9. Shift the directional control valve lever to retract the cylinder rod. Read and record the pressure at each of the gauge positions while the cylinder rod is retracting.
10. When the cylinder is fully retracted, hold the directional control valve lever in the shifted position and record the pressure at each gauge position.
11. Adjust the flow control valve to produce a variety of cylinder extension times (15 seconds, 20 seconds, etc.). Observe the pressure gauge readings during cylinder extension and retraction for each of these settings and note any variations that occur. Time the retraction of the cylinder at each of the settings.
12. Adjust the flow control valve to produce a cylinder extension time of 10 seconds. During the extension and retraction of the cylinder rod, apply a load by using your hands to resist the movement of the rod. Record any variation to gauge readings or movement of the rod.
13. Discuss your data and observations with your instructor before disassembling the system.
14. Disassemble, clean, and return components to their assigned storage locations.

Data and Observation Records

Record pressure readings and general observations concerning the operation of the hydraulic system.

A. Retraction time

Retraction time of the cylinder in seconds: ____________________

B. Pressure readings

System Operation	Gauge Position			
	1	2	3	4
Direction valve centered.				
Cylinder rod extending.				
Cylinder rod extended; valve lever held in the shifted position.				
Cylinder rod retracting.				
Cylinder rod retracted; valve lever held in the shifted position.				

C. General observations

1. Describe any variations observed in pressure readings and general system operation as cylinder extension times are changed during step 11 of the procedure.

2. What variations are observed in pressure readings and general system operation as a load is applied to the cylinder during step 12 of the procedure.

Activity Analysis

Answer the following questions in relation to the collected data and observations of system operation.

1. Why does the gauge at position 1 read 200 psi whenever the cylinder is not moving?

2. What causes the pressure reading on the gauge at position 4?

3. Compare the data collected in steps 7 and 9 of the procedure. Why is the pressure needed to extend the cylinder (gauge at position 2) different than the pressure required to retract it (gauge at position 3)?

4. When the directional control valve is held in the shifted position in step 8 of the procedure, why is the pressure at gauge position 2 roughly the same as the reading at gauge position 1?

____

____

____

____

____

____

5. Describe the pressure variations in the system you encounter in step 11 of the procedure as the extension time of the cylinder was changed. What changes occurred in cylinder *retraction* time?

____

____

____

____

____

6. What happened to the pressure readings at the various gauge positions when a load was applied to the cylinder during step 12 of the procedure? Why?

____

____

____

____

Activity 2-4

Basic Pneumatic System Assembly and Operation

Name ______________________________ Date ______________

In this activity, you will work directly with actual pneumatic system components, including a compressor unit, air maintenance devices, control valves, actuators, and the conductors involved in the operation of a system. The activity familiarizes you with system construction and operation. You will have the opportunity to assemble the system, check compressor unit operating pressure, set workstation operating pressure, adjust actuator operating speed, and control the direction of actuator movement.

Activity Specifications

Construct the basic pneumatic system shown in the following diagram. Operate the system and record observations of the simple tests described in this activity.

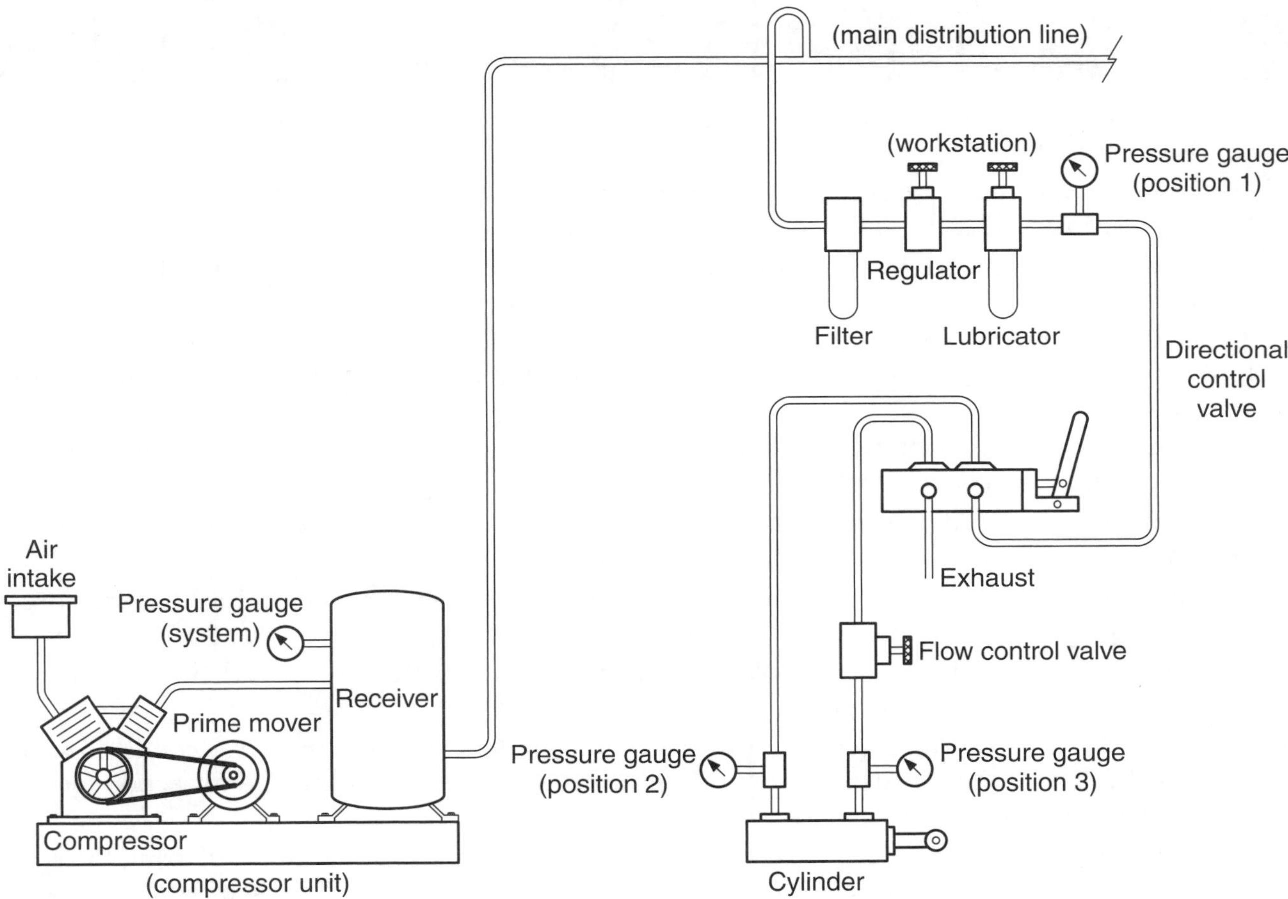

Required Components

Access to the following components is needed to complete this activity.

- Pneumatic compressor unit.
- Pressure regulator.
- Filter.

- Lubricator.
- Directional control valve (four way, two position, manually operated).
- Flow control valve (with return check valve).
- Cylinder (double acting).
- Pressure gauges (four).
- Manifolds (adequate number to allow easy assembly of system).
- Hoses (adequate number to allow assembly of the system).

Procedures

Study the following list of procedures to become familiar with the steps needed to complete the activity. Then, complete the procedure and record the observations indicated in the next section.

1. Select the components needed to assemble the system.
2. Assemble the system and have your instructor check the setup before proceeding.

> **Caution:** Set the pressure regulator for your system to its *lowest* setting before connecting the system to the distribution line connected to the compressor.

3. Measure the system distribution line pressure and record it in Section A of the Data and Observation Records section.
4. Turn on the air to your system and adjust the pressure regulator until the pressure gauge at position 2 reads 40 psi when the cylinder is fully extended.
5. Shift the directional control valve to extend and retract the cylinder to assure the system is operational.
6. Adjust the flow control valve until cylinder extension takes 10 seconds. Time and record the retraction time of the cylinder. Record the time in Section B of the Data and Observation Records section.
7. Shift the directional control valve lever to extend the cylinder rod. Read and record the pressure at each of the gauge positions while the cylinder rod is extending.
8. When the cylinder is fully extended, hold the directional control valve lever in the shifted position and record the pressure at each gauge position.
9. Shift the directional control valve lever to retract the cylinder rod. Read and record the pressure at each of the gauge positions while the cylinder rod is retracting.
10. When the cylinder is fully retracted, hold the directional control valve lever in the shifted position and record the pressure at each gauge position.
11. Adjust the flow control valve to produce a variety of cylinder extension times (5 seconds, 15 seconds, etc.). Observe the pressure gauge readings during cylinder extension and retraction for each of these settings and note any variations that occur.
12. Adjust the flow control valve to produce a cylinder extension time of 10 seconds. During the extension and retraction of the cylinder rod, apply a load by using your hands to resist the movement of the rod. Record any variation to gauge readings or movement of the rod. Also, note any sponginess in cylinder movement.
13. Discuss your data and observations with your instructor before disassembling the system.
14. Disassemble, clean, and return components to their assigned storage location.

Data and Observation Records

Record pressure readings and general observations concerning the operation of the pneumatic system.

A. Line pressure

System distribution line pressure: ______________________________

B. Retraction time

Retraction time of the cylinder in seconds: ______________________________

C. Pressure readings

System Operation	Gauge Position			
	1	2	3	System Pressure
Cylinder rod extending.				
Cylinder rod extended; valve lever held in the shifted position.				
Cylinder rod retracting.				
Cylinder rod retracted; valve lever held in the shifted position.				

D. General observations

1. Describe any variations observed in pressure readings and general operation as cylinder extension times are changed during step 11 of the procedure.

____

____

2. What variations are observed in pressure readings and general system operation as a load is applied to the cylinder during step 12 of the procedure?

____

____

Activity Analysis

Answer the following questions in relation to the data collected and observations of system operation.

1. Why does the pressure in the system distribution line always remain higher than the pressure at the other gauge positions?

____

____

____

2. Why is the retraction time of the cylinder approximately the same for all settings of the flow control valve in step 11 of the procedure?

____

____

____

____

____

3. Why does the pressure gauge reading at position 3 return to 0 psi when the cylinder is fully extended?

4. Why are there differences in the pressures at the various gauge positions as the cylinder extends?

5. Describe the pressure variations in the system as the extension times are changed in step 11 of the procedure. Are the variations most noticeable on extension or retraction? What do you think causes these variations?

6. As the load is applied to the cylinder during step 12 of the procedure, when is movement the most spongy. What do you think causes this sponginess?

Chapter 3

Basic Physical Principles

Applications to Fluid Power Systems

The activities in this lab exercise are designed to show several practical applications of the principles discussed in the chapter. The activities are based on the design and operation of equipment that is usually available in schools, manufacturing facilities, and businesses. Some of the activities include problems that require the application of formulas and appropriate calculations for solution. Guidelines regarding the selection of equipment is given in the introduction to each activity. Permission to observe the equipment *must* be obtained in all cases. Be certain to indicate that all safety regulations required by the company, school, or other institution will be observed.

> **Note:** Check with your instructor to be certain that all the policies of the school are followed, especially if your activities involve a group other than the school.

Key Terms

The following terms are used in this chapter. As you read the text, record the meaning and importance of each. Additionally, you may use other sources, such as manufacturer literature, an encyclopedia, or the Internet, to obtain more information.

absolute pressure ____________________

absolute zero ____________________

alternating current ____________________

atmosphere ____________________

atoms ____________________

basic machine types ____________________

Bernoulli's theorem ____________________

Boyle's law ____________________

Btu ____________________

buoyancy ____________________

calorie ____________________

Charles' law ____________________

circuits ____________________

conduction ____________________

convection

current

degrees

direct current

electrical conductor

electrical insulator

electrical potential

electromagnetic induction

electrons

energy

first law of thermodynamics

first-class lever

flux

force

friction

fulcrum

gauge pressure

Gay-Lussac's law

general gas law

ground

head

heat

heat insulators

horsepower

ideal gas laws

inclined plane

inertia

kinetic energy

latent heat

latent heat of fusion

latent heat of vaporization

lever

lever arm

load arm

magnetic poles

magnetism

mass

mechanical advantage

mechanical efficiency

molecules

neutrons

nucleus

Ohm's law

parallel circuit

Pascal's law

potential

potential energy

power

pressure

protons

pulley

radiation

resistance

screw

second law of thermodynamics

second-class lever

sensible heat ____

____

series circuit ____

____

specific gravity ____

____

specific heat ____

____

specific weight ____

____

standard atmospheric pressure ____

____

temperature ____

____

thermodynamics ____

____

third-class lever ____

____

torque ____

____

torque arm ____

____

vacuum ____

____

velocity ____

____

wedge ____

____

weight ____

____

wheel and axle ____

____

work ____

____

Chapter 3 Quiz

Name ______________________ Date __________ Score ______

Write the best answer to each of the following questions in the blanks provided.

1. List the six simple machines.

____

____

____

2. A lever in which the fulcrum is located between the effort and load arms is called a(n) ______.

____

3. The gears found in many modern machines use the ______ simple machine principle in their operation.

____

4. The energy stored in a form that allows it to be easily released and used is known as ______.

____

5. Energy that is wasted during the operation of an inefficient fluid power system is usually lost in the form of ______ that enters the atmosphere and cannot be ______.

____

6. The effort that ______, ______, or ______ the motion of a body is known as force.

____

7. The common unit designation of pressure in fluid power systems in the United States is ______.

____

8. A force of 50 pounds applied over a distance of 20 feet in 2 minutes produces ______ foot-pounds per minute of power.

____

9. A device that requires an input effort of 25 pounds to lift a weight of 75 pounds has a mechanical advantage of ______.

____

10. Heat is a form of ______ energy.

____

11. British thermal units and calories are used to measure ______.

____

12. The temperature scale that is commonly used in the world outside of the United States is the ______ scale.

____

13. The heat required to produce a change of state in a substance, such as changing from a liquid to a solid, is known as ______ heat.

____

14. The physical movement of a gas or liquid is involved in the ______ mode of heat transfer.

____

15. ______ is the difference between the electron pressure levels of a negatively charged substance and a positively charged substance.

____

16. *Magnetism* is the characteristic of a substance that causes it to attract ______.

17. Electricity can be produced, using the principle of ______, by cutting magnetic lines of force with an electrical conductor.

18. The electrical current that involves electron flow in one direction only and is often used in mobile fluid power systems is known as ______ current.

19. A circuit that provides an individual route for the current to follow through each load is called a(n) ______ circuit.

20. Regarding general characteristics of fluids, what are the two key differences between liquids and gases?

21. Standard atmospheric pressure registers as 0 psi on a gauge calibrated on the ______ pressure scale.

22. Standard atmospheric pressure is equivalent to a column of mercury ______ inches tall.

23. Bernoulli's theorem involves a relationship between fluid pressure, ______, and velocity in a system.

24. Gas law calculations make use of the ______ temperature scale.

25. The general gas law combines the principles involved in ______, ______, and ______ laws.

Activity 3-1

Applications of the Basic Principles of Mechanics

Name ______________________________ **Date** ______________

This activity is designed to show several practical applications of the principles discussed in the mechanics portion of the chapter. The equipment selected for this activity should involve several levers, gears, and other devices for adjustment and operation. The identification and selection of equipment is the responsibility of the student, unless otherwise directed by the instructor.

Activity Specifications

Study the list of questions below to become familiar with the concepts that will be observed. Use this information as a guide to select appropriate equipment to be studied. Identify a machine that appears to contain as many of these factors as possible. If possible, at least three different types of simple machines, including a first-class lever and a wheel and axle, should be involved in the equipment. The device does *not* need to involve fluid power in its operation. Obtain permission to observe the machine as it is operated as well as during downtime. Complete the questions below based on your observations.

1. In general terms, describe the equipment and what it does.

____

____

____

____

2. Where is the equipment located and who is the owner?

____

____

3. List the various applications of simple machines in the equipment. Classify the applications by simple machine type and briefly relate the functions of each type to the operation of the equipment.

____

____

____

____

4. Identify a specific application of a first-class lever in the equipment. Sketch the application showing the dimensions and layout of the effort arm, fulcrum, and load arm. Describe the function of the lever in the overall operation of the equipment.

____

5. Using the information collected in item 4, calculate the mechanical advantage produced by the lever design. Show the formula and calculations used to obtain the answer.

6. Using the information collected in item 4, calculate the resistance that can be overcome if a force of 75 pounds is applied to the end of the effort arm. Show the formula and the calculations used to obtain the answer.

7. Identify a specific application of a wheel and axle simple machine in the equipment selected for this activity. Sketch the application showing the layout and dimensions of the components that function as the simple machine. Describe the function of the wheel and axle principle in the overall operation of the equipment.

____

____

8. Using the information collected in item 7, calculate the mechanical advantage produced by the wheel and axle design. Show the formula and the calculations used to obtain the answer.

9. Using the information collected in item 7, calculate the resistance that can be overcome on the axle portion of the device if a 75 pound force is applied to the outer edge of the wheel. Show the formula and the calculations used to obtain the answer.

10. Identify a third application of a simple machine in the operation of this equipment. This application should involve a simple machine other than a lever or wheel and axle. Sketch the application to show the layout of the device in the equipment. Describe the function of this simple machine in the overall operation of the equipment.

____

____

11. Describe two applications of torque in the maintenance or operation of this piece of equipment. Discuss how the torque is applied and how it can be measured.

____

____

____

____

12. Discuss the concept of *energy* as it applies to this piece of equipment. What is the source of energy for the operation of the equipment? Besides the mechanical means discussed in the previous questions, identify and discuss the methods used to transfer energy in this equipment.

____

____

____

____

Activity 3-2

Applications of the Basic Principles of Heat Transfer

Name ______________________________ **Date** ______________

This activity is designed to show several practical examples of the principles discussed in the heat transfer portion of the chapter. The activity involves heat transfer principles as they relate to all aspects of selected equipment, but places special emphasis on the fluid power system. The equipment selected for this activity should involve a fluid power system, including the hydraulic pump or air compressor unit. The identification and selection of equipment is the responsibility of the student, unless otherwise directed by the instructor.

Activity Specifications

Study the list of questions below to become familiar with the concepts that will be observed. Use this information as a guide to select appropriate equipment to be studied. Identify a machine that appears to contain as many of these factors as possible. The device *should* involve fluid power in order to emphasize heat transfer principle as they apply to fluid power equipment. Obtain permission to observe the equipment during all phases of machine operation. Complete the questions below based on your observations.

1. In general terms, describe the equipment and what it does.

____
____
____
____

2. Where is the equipment located and who is the owner?

____
____

3. Why is heat generated by the power unit of the fluid power system? Relate this to the energy input from the prime mover of the system and the energy output of the actuator(s). Identify the component that appears to generate the most heat. Why does heat generation occur in this component?

____
____
____
____
____
____
____
____

4. Is the heat generated by the power unit sensible or latent heat? Explain your answer.

____
____
____
____

5. Measure and record the air temperature in the room with a Fahrenheit-scale thermometer. Convert the temperature to the Celsius scale. Show all of your calculations.

6. Convert the Fahrenheit room temperature measured in item 5 above to the Rankine absolute scale. Show all of your calculations.

7. Identify an example of heat transfer by conduction in the equipment. Describe how conduction operates in this specific application to help maintain the selected operating temperature.

8. Identify an example of heat transfer by convection in the equipment. Describe how convection is used to help maintain the proper operating temperature of the equipment. How important is convection heat transfer in the cooling of the equipment?

9. How important is the transfer of heat by radiation in the operation of the equipment? Explain your answer, including a discussion of both the application and the extent of use of heat radiation.

10. Discuss the term *specific heat* and why individuals working in the fluid power area should be familiar with its definition. Using reference publications in the fluid power laboratory, the school's library, or trusted Internet sources, identify the specific heat of:

A. Aluminum ______________________________

B. Cast iron ______________________________

C. Hydraulic oil ______________________________

D. Polystyrene plastic ______________________________

11. Discuss the principle of the *conservation of energy* as it relates to the equipment. What has happened to all of the energy used by the prime mover in this system?

____

____

____

____

12. Discuss in general terms the second law of thermodynamics. Why is this law so important in maintaining proper operating temperature of equipment?

____

____

____

____

Activity 3-3

Applications of the Basic Principles of Electricity and Magnetism

Name ______________________________ **Date** ______________

This activity is designed to show several practical examples of the principles discussed in the electricity and magnetism portion of the chapter. The equipment selected for this activity should involve a fluid power system that *includes* electrical control of the prime mover and at least a part of the components that are included in the fluid power system circuit(s).

Activity Specifications

Study the list of questions below to become familiar with the concepts that will be observed. Use this information as a guide to select appropriate equipment to be studied. The equipment should involve fluid power in order to emphasize how the principles of electricity and magnetism are basic to the operation and control of the prime mover and control portion of both hydraulic and pneumatic systems. Obtain permission to observe the equipment during all phases of machine operation. Complete the following questions based on your observations.

Caution: Be certain to disconnect all electrical equipment or have appropriate fuses removed before examining control boxes and other electrical devices.

1. In general terms, describe the equipment and what it does.

____

____

____

____

2. Where is the equipment located and who is the owner?

____

____

3. Describe the prime mover for the equipment. What electrical devices are used to control this source of power for the system? In general terms, describe the devices.

____

____

____

____

4. Describe at least two locations where magnetic fields are involved in the operation of the equipment. Describe the source of the magnetic lines of force at each of the locations.

____

____

____

____

5. Examine the data plate of the prime mover to obtain the following information:
   A. Current type of the unit (ac or dc) ______________________
   B. Voltage of the unit ______________________
   C. Cycles per second of operation ______________________
   D. Amperage of the unit ______________________
6. Identify at least two locations where solenoids are used in the equipment. What is the purpose of the solenoid at each of these locations?
7. Describe the various functions in the system that use alternating current.
8. Determine if direct current is used in any part of the equipment. Describe any applications that are found in the system and provide a rationale for the use of direct current, rather than alternating current, in that portion of the equipment.
9. Using the Ohm's law formula, calculate the amperage required in a 120V system to operate when a 50Ω resistance is encountered. Show the formula used and the calculations to determine the current needed.

10. Locate the four basic electrical circuit elements in the prime mover control circuit or one of the general control circuits in the equipment. Sketch the circuit elements and label the components with their name and function. Indicate the route of electrical flow through the circuit.

11. Locate a diagram of the electrical circuit that controls the fluid power control valve portion of the equipment. Study the drawing and locate a portion of the diagram that you feel illustrates a series connection in the circuit. Make a sketch of that portion of the diagram. Indicate why the circuit portion is a series connection and why a series design may have been used in that part of the circuit.

____

____

____

____

12. Further study the diagram of the electrical circuit identified in item 11 to locate a portion of the circuit that you feel illustrates a parallel connection in the circuit. Make a sketch of that portion of the diagram. Indicate why the circuit portion is a parallel connection and why a parallel design may have been used in that part of the circuit.

____

____

____

____

Activity 3-4

Applications of the Basic Principles of Fluid Power Transmission

Name ________________________________ **Date** ______________

This activity is designed to show several practical examples of the principles discussed in the fluid power transmission portion of the chapter. Access to equipment that uses both hydraulics and pneumatics is required to complete the activity. If possible, the pneumatic system should include a compressor and receiver in addition to the actuators and control valves. If equipment is only available that operates from a central air supply, access to the central compressor and receiver is necessary.

Activity Specifications

Study the list of questions below to become familiar with the concepts that will be observed. Use this information as a guide to select equipment with appropriate components and circuits. It is recommended that two pieces of equipment be identified for use in this activity: one primarily hydraulic powered and one primarily pneumatic powered. Complete the questions below based on your observations.

1. In general terms, describe the equipment and what it does. If two pieces of equipment are being observed, describe both.

____

____

____

____

____

____

2. Where is the equipment located and who is the owner?

____

____

3. Observe the operation of the hydraulic and pneumatic equipment. Note the differences in the two systems on factors such as noise level, speed of operation, and cleanliness. Discuss how the observed differences relate to the basic characteristics of the fluids used to operate the systems.

____

____

____

____

____

____

4. Discuss the term *specific weight*. How does it differ from specific gravity? Using reference publications from the fluid power laboratory, school library, or trusted Internet sources, identify the specific weight of the following materials. Why does the value vary so much between the materials?

____

   A. Aluminum ______________________________
   B. Polystyrene plastic ______________________________
   C. Cast iron ______________________________
   D. Air ______________________________
   E. Hydraulic oil ______________________________
   F. Oxygen ______________________________

5. Obtain a sample of hydraulic oil and test the specific gravity using a hydrometer. Record both the specific gravity and the temperature of the oil. Why must the specific gravity value be adjusted for temperature?

____

____

6. What is the maximum operating pressure shown on the gauges of the equipment using the hydraulic system? Which pressure scale is used on these gauges? How did you determine which scale is in use?

____

____

____

____

7. Convert the operating pressure identified in item 6 to the scales shown below. Show the calculations used to make these conversions.

   A. Gauge pressure ______________________________
   B. Absolute pressure ______________________________
   C. Head of water______________________________

8. Convert a gauge pressure reading of –2.5 psig to the vacuum scale. Express your answer in inches of mercury. Show the calculations used to make this conversion.

____

9. Observe the layout of the tubing and components of the pneumatic system used in this activity. Apply the basic concept of Bernoulli's theorem to the airflow through the system. Sketch a section of a circuit used in the equipment where you feel the principle involved in the theorem causes air velocity and pressure changes. Discuss what would happen at these points and why changes should occur.

____

____

____

____

10. Measure or estimate the diameter of the largest actuator in the systems you observed for this activity. What force could the actuator exert if the system is operating at 100 psig? Show all of your calculations. Which basic fluid power principle is involved in this calculation? Does this principle apply the same to both hydraulic and pneumatic systems? Why or why not?

____

____

____

____

11. Establish the volume of the pneumatic system receiver used with the equipment selected for use with this activity. This may be obtained from manufacturers specifications or by calculations using the length and diameter of the tank. Calculate the volume of air at atmospheric pressure that is stored in the receiver at the maximum pressure setting of the compressor. Ignore temperature differences.

____

12. Calculate the pressure change in the receiver in item 11 when the air is allowed to cool to 80°F from an initial pressure of 100 psig and a temperature of 95°F. What implications does your answer have for the operation of pneumatic fluid power systems?

____

____

Fluid Power Standards and Symbols

Language of the Industry

The following activities are designed to show several practical applications of the principles discussed in this chapter. The activities are based on the study of hydraulic and pneumatic standards, symbols, and diagrams. Each activity is designed to provide experience with the general principles discussed in the chapter. These principles include the structure of a standard and the use of symbols representing actual components and circuits of fluid power systems.

Key Terms

The following terms are used in this chapter. As you read the text, record the meaning and importance of each. Additionally, you may use other sources, such as manufacturer literature, an encyclopedia, or the Internet, to obtain more information.

accumulators ______________________________

air receiver ______________________________

American National Standards Institute (ANSI) ______________________________

American Petroleum Institute (API) ______________________________

ASME ______________________________

ASTM International ______________________________

circuit diagrams ______________________________

control mechanisms ______________________________

cutaway symbols ______________________________

de facto standard ______________________________

due process ______________________________

electrical controls ______________________________

energy conversion devices ______________________________

feedback controls ______________________________

Fluid Power Distributors Association (FPDA) ______________________________

Fluid Power Educational Foundation (FPEF) ______________________________

general consent standard ______________________________

graphic symbols ______________________________

International Electrotechnical Commission (IEC)

International Fluid Power Society (IFPS)

International Organization for Standardization (ISO)

legal regulation standard

mechanical controls

muscular controls

National Fluid Power Association (NFPA)

pictorial symbols

pressure controls

standard

symbols

Underwriters Laboratories, Inc. (UL)

Chapter 4 Quiz

Name ______________________ **Date** __________ **Score** ______

Write the best answer to each of the following questions in the blanks provided.

1. Define *standard.*

2. It is fairly common in international business to use standards based on the ______ development process.

3. Name the six general classifications of groups that are involved in standards development.

4. Standards developed by ______ are often considered to best reflect the needs of the field because they are developed by individuals working in the field.

5. ______ of organizations with diverse backgrounds usually develop standards for technologies that combine several different fields, such as for the volatile information and communication fields.

6. The principle coordinating group that works with standards on the international level is the ______.

7. Testing and certifying groups, such as Underwriters Laboratories, Inc., develop standards and test the safety of component and equipment ______, ______, ______, and ______.

8. The fluid power organization that provides members training programs on topics such as product distribution management, quality processes, and sales management is the ______.

9. The Fluid Power Educational Foundation is an organization with the primary objective of supporting and promoting ______ in the fluid power field.

10. During the development of a standard, a carefully outlined set of procedures is followed for the ______, revision, ______, and withdrawal phases.

11. The content in a standard is typically organized into sections. List the names of the seven sections commonly found in a standard.

12. The most common type of fluid power symbols, which are symbols made up of a series of lines and standard figures, are called ______ symbols.

13. The symbol type that is the most standardized is the ______ symbol.

14. Symbols that show the exterior shape of components and the connecting lines are known as ______ symbols.

15. Symbols are shown in the ______ position, unless they are part of a circuit diagram showing a sequence of operation.

16. Working, pilot supply, and return fluid power lines are all represented by a ______ linetype in graphic symbols.

17. In a diagram using graphic symbols, the largest circles represent components such as ______, ______, and ______.

18. The primary components represented by a rectangle are ______ and ______.

19. The junction of lines is indicated by a(n) ______ placed at the point where the lines meet.

20. List the five classifications of control mechanisms for which symbols are available.

21. The basic symbol for an accumulator is a(n) ______ with operational details shown by additional symbols on the inside of the symbol.

22. The arrow that shows flow through a valve does not align with the external ports of the valve in the symbol for a normally ______ pressure control valve.

23. The simplified symbol for a flow control valve indicates that the valve is ______ if an arrow at a 45° angle appears across the restrictive orifice of the valve.

____

24. The basic graphic figure used to construct a measuring instrument symbol is a ______.

____

25. List eight items that should be included in circuit diagrams for fluid power systems developed according to the ISO standard 1219-2.

____

____

____

____

____

____

____

____

Activity 4-1

Analysis of the Structure and Technical Content of a Fluid Power Standard

Name ______________________________ **Date** ______________

This activity provides an opportunity to examine an actual fluid power standard. Standards of this type are used to allow manufacturers in this country and around the world to design and produce components with similar purposes. Examining standards such as the one selected for this activity illustrates how internationalization of components and products has been made possible.

Activity Specifications

The document used in this activity should be a published ANSI, NFPA, or ISO standard. Obtain a printed or Internet-based copy of a standard from your instructor. Carefully read the document and then answer the following questions.

1. Describe the scope of this standard.

____

____

____

____

2. What is the purpose of this standard?

____

____

____

____

3. What system components are described in this standard?

____

____

____

____

4. Are operating pressure, flow rates, and fluid temperature a part of this standard? If they are included, explain why and if not, explain why not.

____

____

____

____

5. What size of lines (inlet, outlet, drain, pilot) are recommended by this standard?

____

____

____

____

6. Are specifications regarding pressure drop through the component included in this standard? If a specific figure is not included, in your opinion, why is a pressure drop limit not a factor for this standard?

7. Identify the sections of the standard that specifically deal with safety aspects of the component or system.

8. What level of component or system fluid filtration is recommended by this standard?

9. Describe any maintenance factors included in the standard. What is the general thrust of these factors?

10. In your opinion, how complete is this standard? Does the standard provide the information that a fluid power specialist needs to set up, operate, and maintain a fluid system?

Activity 4-2

Identification of Fluid Power Components and Function from Graphic Symbol Circuit Drawings

Name ______________________________ **Date** ______________

This activity provides an opportunity to identify components and their characteristics from the graphic symbols used in a typical fluid power circuit drawing. Manufacturers often use variations of standardized symbols to illustrate the operation of their systems. In addition, word descriptions are often used to explain the sequential steps of circuit operation.

Activity Specifications

Study the following graphic symbol circuit drawing and the description of the operation of that circuit and the machine it powers. Identify the component parts and their function using the circuit drawing and the system description. Use the information provided in the text and other available references about the symbols involved in the circuit. Then, complete the activity questions based on your analysis of the circuit.

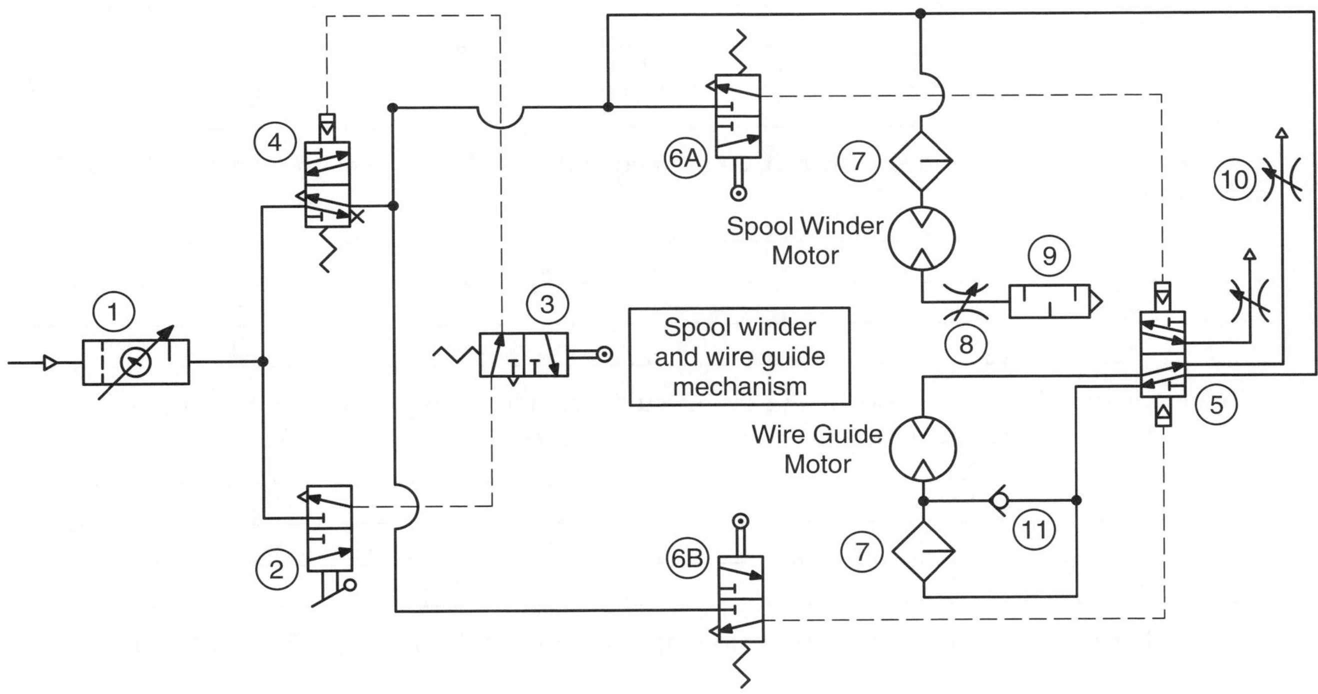

System Description

The device is a small, wire-winding machine designed to transfer electrical wire from a large bulk spool to smaller spools. The machine operator mounts the large, wire-supply spool on one end of the machine, runs the free end of the supply wire through a guide mechanism, and attaches the wire to a smaller take-up spool. The manually operated valve at location 2 is shifted to supply pilot pressure to valve 4 through valve 3. Valve 3 serves to limit the amount of wire wound on the take-up spool. Fluid to operate and control both the winding and guide motors is supplied through valve 4.

The wire guide motor operates a guide mechanism to assure a smooth, even wind of the supply wire on the smaller take-up spool. As the guide mechanism moves from one end of the take-up spool to the other, it trips a limit valve at location 6A or 6B. This changes the rotational direction of the guide motor by switching control valve 5. As the wire take-up spool becomes full, it trips valve 3 removing the pilot pressure from supply valve 4, which stops both motors. Once this occurs, the machine operator manually returns valve 2 to the exhaust position.

The rate of movement of the guide mechanism is set by valves 10 at the outlet of valve 5. Valve 8 controls the speed of the winding motor.

Activity Questions

1. Which fluid is used to operate this system? Indicate specifically how you are certain that this is the fluid type.

____

____

2. What is the function of component 1 in this circuit?

____

____

____

____

3. What is the function of component 9 in this circuit? Why do you think this component is used only in conjunction with the spool winder motor?

____

____

____

____

4. Provide a complete name description of component 2 that both describes the construction of the valve and its function.

____

____

____

____

5. What is the function of component 3 in this circuit? Provide a complete name description that includes both the construction of the valve and its function.

____

____

____

____

6. What shifts component 4 when the wire take-up spool fills and component 3 is shifted by the filled spool?

____

____

____

____

7. Which feature of the graphic symbols for the guide and winder motors indicate that the motors are designed to operate in either direction?

____

____

8. What is the function of the components numbered 7 in this circuit?

9. Describe the design features of components 6A and 6B in the circuit. What is the function of the components in this circuit?

10. What is the name of component 11 in this circuit? Why is the component used in that circuit location?

Activity 4-3

Creation of Fluid Power Graphic Symbols from Specific Component and System Characteristics

Name ____ **Date** _______________

This activity provides an opportunity to draw graphic fluid power symbols from a description and pictorial diagram of a circuit. Manufacturers sometimes use combinations of the different symbol styles when illustrating circuit operation. For example, various forms of pictorial symbols are often found in advertising brochures or in training information to show the relationship of components, rather than specific details.

Activity Specifications

Study the pictorial diagram and carefully read the circuit description before completing the questions. Complete the questions by providing a graphic symbol for the designated component. Be certain to include only the symbol detail justified by the descriptions provided in the pictorial diagram and circuit description. Complete the questions using simplified symbols whenever possible. Use the information provided about symbols from the text and other available references about symbols. Additional questions require an analysis of the pictorial circuit diagram and circuit description.

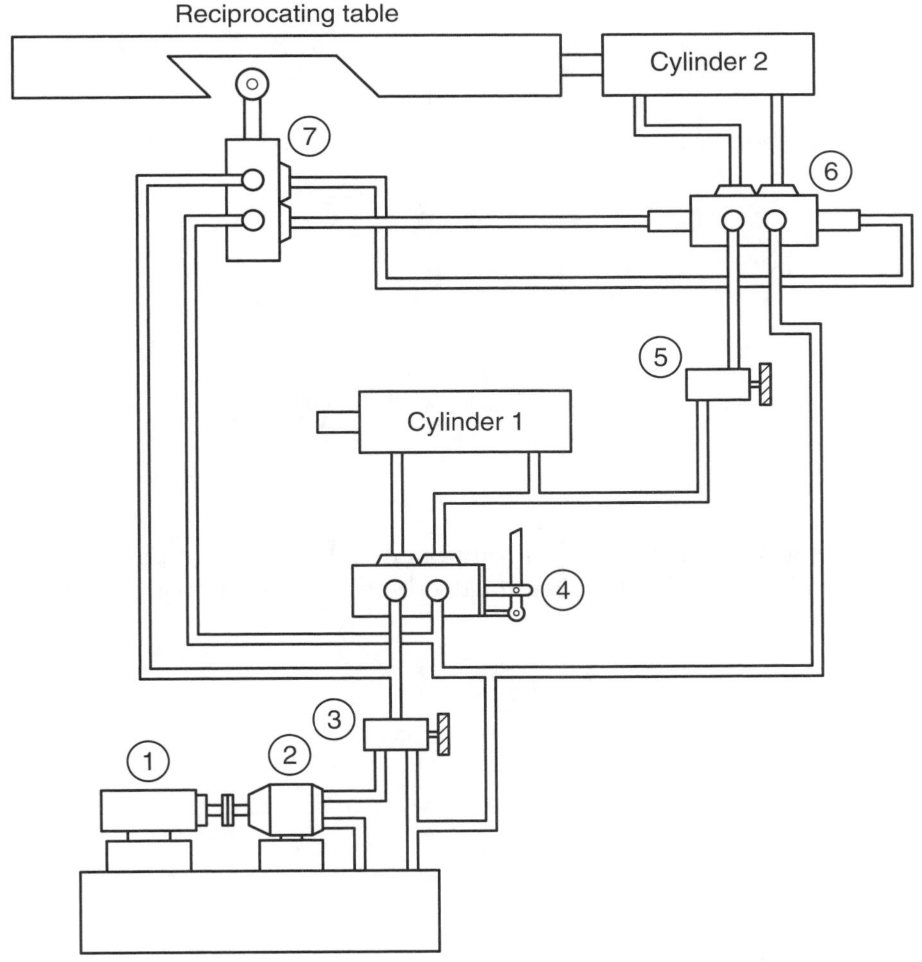

Circuit Description

This is a clamp and reciprocation circuit used on machine tools where a workpiece must be clamped before automatic reciprocation of the worktable begins. Shifting the two-position, four-way directional control valve 4 activates cylinder 1, which clamps the workpiece. When the clamp is closed, system pressure builds to the setting of the sequence valve 5.

When the sequence valve 5 opens, fluid is allowed to move through valve 6. Valve 6 is a pilot-operated, two-position, four-way directional control valve that operates cylinder 2, which powers the reciprocating worktable. When the table reaches the forward or reverse end of its stroke, it activates valve 7. Valve 7 is a two-position, four-way directional control valve shifted by a cam on the table. Shifting valve 7 directs pilot pressure to the opposite side of valve 6, which shifts and reverses the movement of the table.

Reciprocation continues until valve 4 is shifted to release the clamping pressure on the workpiece. Maximum system pressure is established by the relief valve 3.

Activity Questions

1. Draw the graphic symbol for the prime mover shown as component 1 on the pictorial diagram.

2. Draw the graphic symbol for the pump shown as component 2 on the pictorial diagram.

3. The relief valve for the system is shown as component 3. Draw the graphic symbol for that valve providing in the symbol only the details indicated by the pictorial diagram and in the circuit description.

4. What is the primary function of component 4 in the pictorial diagram? Draw the graphic symbol for that valve including the activation method.

____

____

____

____

5. What is the function of component 5 in the pictorial diagram? Draw the graphic symbol for the valve. Be certain to show the proper drain type for this valve.

____

____

____

____

6. What is the function of component 6 in the pictorial diagram? Draw the graphic symbol for this valve. How is this valve shifted?

____

____

____

____

7. What is the function of component 7 in the pictorial diagram? Draw the graphic symbol for this valve. How is this valve shifted?

8. Draw the graphic symbol for cylinder 1. Is the symbol for cylinder 2 different? Why or why not? If the symbol is different, draw it as well.

9. Draw the line pattern used in graphic symbols to show the function of the lines between components 6 and 7. What are these lines called?

10. What type of lines are between components 4, 6, 7 and the reservoir? How are these drawn in a diagram using graphic symbols?

11. What type of lines are between components 4 and 5, components 5 and 6, and component 6 and cylinder 2? How are these to be drawn in a diagram using graphic symbols?

12. *(extra credit problem)* Convert the circuit shown in the pictorial diagram to a diagram using graphic symbols for all component parts.

Chapter 5

Safety and Health

Promoting Proper Practices

The following activities are designed to show several practical applications of the principles discussed in this chapter. The activities are based on the study of safety and health issues that relate to the fluid power area. They are designed to provide exposure to a number of factors discussed in the chapter. These factors involve the identification of health, safety, and general shop procedures that affect the safety of personnel. Be certain to obtain permission to make observations in laboratories and any businesses you may visit as you conduct your work.

> **Note:** Check with your instructor to be certain that all the policies of the school are followed, especially if your activities involve a group other than the school.

Key Terms

The following terms are used in this chapter. As you read the text, record the meaning and importance of each. Additionally, you may use other sources, such as manufacturer literature, an encyclopedia, or the Internet, to obtain more information.

air-line respirators ______________________

back injuries ______________________

cumulative injuries ______________________

dust mask ______________________

earmuffs ______________________

earplugs ______________________

emergency showers ______________________

equipment maintenance ______________________

eye protection devices ______________________

eyewash stations ______________________

first aid kit ______________________

hearing-protection devices ______________________

housekeeping ______________________

lockout devices ______________________

Occupational Safety and Health Administration (OSHA) ______________________

pressure safety valve ______________________________
____
____

safety shoes ______________________________
____
____

respirator face masks ______________________________
____
____

self-contained breathing devices ______________________________
____
____

safety helmets ______________________________
____
____

training ______________________________
____
____

Chapter 5 Quiz

Name ______________________ **Date** __________ **Score** ______

Write the best answer to each of the following questions in the blanks provided.

1. Injuries that result from long term exposure to unsafe environmental conditions are classified as ______.
____

2. Name three environmental conditions from a shop setting that can cause the injuries described in question 1.
____
____
____

3. The two basic goals of any health and safety training program should be to eliminate ______ and reduce ______.
____
____

4. Appropriate air exchange and circulation in a work environment are important for the maintenance of suitable ______ and ______ levels.
____
____

5. It is the responsibility of the ______ to provide training sessions so a worker has the opportunity to practice the procedures before actually attempting to do the job.
____

6. List five personal factors that can cause an individual worker to perform a job unsafely.
____
____
____
____
____

7. Which of the following benefits can an employer receive if they have a well-established safety and health program in operation?
   A. Decreased benefit and insurance costs.
   B. Reduced employee work time loss caused by sickness and injury.
   C. Increased employee morale.
   D. All of the above.
____

8. What are the three factors relating to the design and construction of safety glasses that ANSI standards control?
____
____
____

9. Safety shoes that can be used to protect the worker include the following three designs. Provide the type of protection each offers.
   A. Steel-toed shoes
   ______
   B. Metal-free shoes
   ______
   C. Wood-soled shoes
   ______
10. Define *housekeeping.*
   ______
   ______
   ______
   ______
11. What is the purpose of a guard?
   ______
   ______
   ______
   ______
12. ______ not only means the service work done on large stationary pieces of equipment, but also maintenance of portable tools and hand tools.
   ______
13. What represents the largest single group of injuries for which compensation is awarded?
   ______
14. All fluid power systems should always be treated as if they are ______ and ready to function.
   ______
15. The fluid power component type that is generally considered the most dangerous is the ______.
   ______
16. The two elements that are considered the most dangerous in the operation of pumps and compressors are the ______ involved in the operation of the component and the ______ located between the prime mover and the component.
   ______
   ______
17. List two factors an air receiver contributes to the safe operation of a pneumatic system.
   ______
   ______
18. List three safety-related rules concerning the location of machine controls on equipment involving fluid power components.
   ______
   ______
19. Workers often have poor safety habits when working with portable, ______ because of the compact and lightweight designs used in many of the units.
   ______

20. Name four factors controlled by the Occupational Safety and Health Administration (OSHA) that influence workers in an industrial setting.

____

____

____

____

Activity 5-1

Facility Health and Safety Assessment

Name ______________________________ **Date** ______________

This activity is designed to show how several of the health and safety issues discussed in the text are implemented in an actual shop operation. The activity may be conducted as an individual project or a class field trip may be planned to provide general observation of a manufacturing facility. When requesting permission to tour a shop, it should be indicated that you will be observing the overall plant operation and that all regulations required by the company will be followed. Be certain to follow all of the policies of your school regarding activities outside the institution.

Activity Specifications

Tour an industrial plant to observe the type of health and safety factors required in the operation of a manufacturing facility. The facility should be large enough to involve a variety of equipment and several workers in its operation. Ideally, the shop will include machines that use fluid power in their operation. Complete the following activity questions based on the plant visit.

1. Describe the physical facilities that house the organization.

____

____

____

____

2. Describe the type of activities done in the facility.

____

____

____

____

3. Describe your impression of the physical layout of the facility. Include information on such items as the adequacy of aisles, work space, machine arrangement, and general atmosphere.

____

____

____

____

4. What is your impression of the adequacy of the lighting for the type of work being done in the facility? Describe the overall light level and any provisions for increased lighting in areas where detailed work is done.

____

____

____

____

5. What provisions appear to be in place to assure quality air in the work areas? Describe any exhaust fans used to eliminate dust, vapors, or gases from specialized work areas. Describe any air intake systems that appear to bring outside air into the facility.

6. Describe any work areas where respiratory devices are being used to protect workers from dust, vapors, or gases. Why do you think these devices are used, rather than increasing overall ventilation in the facility?

7. Are there areas in the facility that require the use of safety helmets (hardhats)? What appears to be the reason for this requirement?

8. Are there areas in the facility that requires the use of safety glasses? What appears to be the reason for this requirement?

9. What signs are visible providing directions in case of fire or natural disaster. Describe the provisions made for each of these events.

10. What type of fire extinguishers are located in the facility? Are these extinguishers the type needed for the fires that might occur in the areas where they are located?

11. Do any jobs in the facility require the use of protective clothing? What are the jobs that require this type of protection? Describe the clothing and the type of protection provided to the employee.

12. What type of guards are used to prevent workers from making contact with dangerous machine parts? Describe at least two different types of guards that are used in the facility.

____

____

____

____

13. Is a program that encourages proper lifting techniques to prevent back injuries evident in any part of the facility? Describe any evidence, such as signs, used to support this health and safety issue.

____

____

____

____

14. Describe any safety or health program directed specifically at hydraulic or pneumatic applications in the facility.

____

____

____

____

15. If fluid power is used in the facility, what type of safety devices are used on the hydraulic power units and pneumatic compressors to prevent excessive pressures in the systems? Describe one system that is used in the facility.

____

____

____

____

16. What is your overall impression of the health and safety aspects of the facility you visited? In your opinion, what are the strong points of the program? What are the weak points?

____

____

____

____

Activity 5-2

Operation of a Basic, Two-Hand-Control Safety Circuit

Name ______________________________ **Date** ______________

This activity provides an opportunity to work with pneumatic system components. The operation of a safety circuit that requires both of the operator's hands on the controls for the equipment to function will be observed. This technique is often used on industrial presses to prevent hand injuries.

Activity Specifications

Construct, operate, and record observations of the pneumatically operated, two-hand-control safety circuit shown in the following diagram. You will have an opportunity to evaluate the safety features of the circuit.

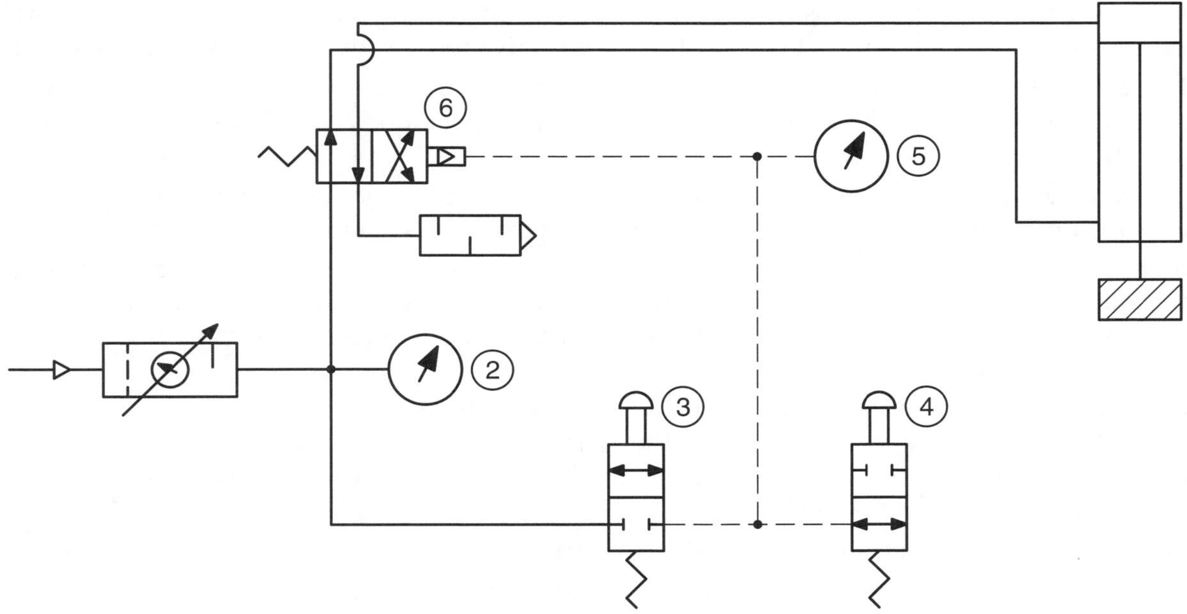

Required Components

Access to the following pneumatic components is needed to complete this activity.

- Pneumatic compressor or building air supply.
- Pressure regulator.
- Filter.
- Lubricator.
- Pressure gauges.
- Directional control valve (four way, two position, pilot operated with spring offset).
- Directional control valve (two way, two position, normally closed with spring offset).
- Directional control valve (two way, two position, normally open with spring offset).
- Cylinder (double acting).
- Manifolds (adequate number to allow easy assembly of system).
- Hoses (adequate supply to allow assembly of system).

Procedures

Study the following list of procedures to become familiar with the steps needed to complete the activities. Then, complete the procedure and record the observations indicated in the next section.

1. Select the components needed to assemble the system.
2. Assemble the system and have your instructor check the setup before proceeding.

Caution: Set the pressure regulator for the system to its *lowest* setting before connecting the system to the distribution line connected to the compressor.

3. Connect your system to the distribution line coming from the air supply and adjust the regulator until the gauge at position 2 reads 30 psi.
4. Shift valve 3 *only* and hold it in the depressed position. Observe and record the motion of the cylinder, the pressure at gauge positions 2 and 5, and any air venting from the system. Release and depress the valve several times to check if repeated activation causes variations in circuit operation.
5. Shift valve 4 *only* and hold it in the depressed position. Observe and record the motion of the cylinder, the pressure at gauge positions 2 and 5, and any air venting from the system. Release and depress the valve several times to check if repeated activation causes variations in circuit operation.
6. Shift valves 3 and 4 simultaneously, holding both valves in the depressed position. Observe and record the motion of the cylinder, the pressure at gauge positions 2 and 5, and any air venting from the system.
7. Release valves 3 and 4 simultaneously. Observe and record the motion of the cylinder, the pressure at gauge positions 2 and 5, and any air venting from the system.
8. Depress and hold down valve 3 to simulate a machine operator taping the valve down to override the safety system. Observe the motion of the cylinder, the pressure at gauge positions 2 and 5, and any air venting from the system. Depress valve 4 and make your observations again. Release valve 4 and repeat your observations.
9. Depress and hold down valve 4 to simulate a machine operator taping the valve down to override the safety system. Observe the motion of the cylinder, the pressure at gauge positions 2 and 5, and any air venting from the system. Depress valve 3 and make your observations again. Release valve 3 and repeat your observations.
10. Discuss your data and observations with your instructor before disassembling the system.
11. Disassemble, clean, and return components to their assigned storage location.

Data and Observation Records

Record pressure readings, cylinder movement, and air venting observations concerning the circuit.

A. Pressure readings, cylinder movement, and air venting observations.

Control Valve Position	Pressure Readings		Cylinder Movement	Air Venting
	Position 2	Position 3		
Valve 3 held in depressed position				
Valve 4 held in depressed position				
Valves 3 and 4 shifted simultaneously				
Valves 3 and 4 released simultaneously				
Valve 3 taped and valve 4 operated				
Valve 4 taped and valve 3 operated				

B. General observations

1. Describe the normal operation of the cylinder when both hand control valves are released.

2. Describe the operation of the cylinder when the valves are taped down to simulate attempted system override.

3. What was observed at valve 4 when valve 3 is taped down?

Activity Analysis

Answer the following questions in relation to the collected data and observation of system operation.

1. Why must both of the hand-operated control valves be simultaneously depressed in order to extend the cylinder? Trace airflow through the circuit to explain your answer.

2. Why does the cylinder always return to the retracted position in this version of the two-handed safety system?

3. What happened to the pressure at gauge position 5 when valve 4 is taped down? How did this affect the operation of the circuit?

4. Why does the cylinder not properly retract when valve 4 is taped down?

5. How effective would this basic safety system be in keeping an operator's hands away from the moving cylinder? Could an operator effectively bypass the system by taping down one of the valves? Explain your answer.

Chapter 6

Hydraulic Fluid

The Energy Transmitting Medium

The activities in this lab exercise are designed to provide experience with actual hydraulic fluid specification sheets issued by fluid producers. The information includes several of the fluid types that are discussed in the chapter. This material is typically available in printed or electronic forms. Completing the activity requires identifying fluid types and making an analysis of the factors that are used to rate fluids.

Key Terms

The following terms are used in this chapter. As you read the text, record the meaning and importance of each. Additionally, you may use other sources, such as manufacturer literature, an encyclopedia, or the Internet, to obtain more information.

95/5 fluids ______

absolute viscosity ______

additives ______

antifoaming agents ______

anti-wear agents ______

API gravity ______

biodegradable fluids ______

capillary viscometer ______

catalysts ______

corrosion inhibitors ______

demulsifier additives ______

emulsion ______

extreme-pressure agents ______

film ______

fire point ______

flash point ______

fluid ______

Four-Ball Method ______

friction ____________________

____

____

HWCF ____________________

____

____

inverted emulsion ____________________

____

____

kinematic viscosity ____________________

____

____

lubrication ____________________

____

____

lubricity ____________________

____

____

oil-in-water emulsion ____________________

____

____

oxidation ____________________

____

____

oxidation inhibitors ____________________

____

____

phosphate ester ____________________

____

____

piston rod wiper rings ____________________

____

____

polyglycol ____________________

____

____

pour point ____________________

____

____

pour-point depressants ____________________

____

____

rotating drum viscometer ____________________

____

____

rust inhibitors ____________________

____

____

Saybolt viscometer ____________________

____

____

sludge ____________________

____

____

specific gravity ____________________

____

____

spontaneous ignition ____________________

____

____

Timken Method ____________________

____

____

viscosity ____________________

____

____

viscosity grades ____________________

____

____

viscosity index number ____________________

____

____

viscosity-index-improver additives ____________________

____

____

Chapter 6 Quiz

Name ______________________ **Date** __________ **Score** ______

Write the best answer to each of the following questions in the blanks provided.

1. Name five functions that the fluid in a hydraulic circuit must perform in addition to transmitting energy to assure an effectively operating system.

2. The friction between two surfaces can easily be reduced by placing a layer of liquid, called a(n) ______, between them.

3. Define *viscosity.*

4. Name three major problems that may appear if a fluid with too high of a viscosity is used in a hydraulic system.

5. The pour point of a hydraulic fluid is especially important when a fluid power system is operated in very ______ weather conditions.

6. The temperature of the fluid in the reservoir of an operating fluid power system is ideally between ______ and ______ degrees Fahrenheit.

7. The metal used in fluid power systems that is generally recognized as the greatest promoter of fluid deterioration is ______.

8. Excessive air entrainment and foaming may be caused by a variety of hydraulic system design flaws or operational problems. Name three typical factors that can cause these problems.

9. The temperature at which a fluid is vaporizing rapidly enough to continuously burn after a flame is applied is called the ______.

10. A primary disadvantage of petroleum-based hydraulic fluids is low resistance to ______, which makes them unsuitable in many applications.

11. Hydraulic fluids that use soluble oil to form an oil-in-water emulsion generally have ______ percent or less oil in their formulation.

____

12. Name three problems that are commonly associated with the use of fire-resistant fluids in hydraulic systems.

____

____

____

13. The water content of water-in-oil emulsions is approximately ______ percent, which is considerably lower than both the soluble-oil emulsion and the high-water-content fluids.

____

14. Which fire-resistant fluid is similar to automotive antifreeze?

____

15. Synthetic fire-resistant hydraulic fluids have several advantages over other fire-resistant fluids. Name four of these advantages.

____

____

____

____

16. ______ agents prevent metal-to-metal contact by forming protective coatings on the internal parts of components.

____

17. The additive that allows a hydraulic fluid to flow more freely at low temperatures is known as a(n) ______ depressant.

____

18. Decreasing ______ allows antifoaming agents to reduce the formation of bubbles and foam in a hydraulic system reservoir.

____

19. ______ is also called dynamic viscosity.

____

20. Kinematic viscosity is measured with the aid of a device called a(n) ______.

____

21. The ISO viscosity rating system groups fluids into 20 grades, each of which may have a kinematic viscosity variation of plus or minus ______ percent from the stated grade number.

____

22. The system that indicates hydraulic fluid viscosity by the time it takes 60 ml of the fluid to flow through a calibrated orifice is called the ______ system.

____

23. Blowing air through a hydraulic fluid at a measured rate and given time period is used to determine the ______ characteristics of the material.

____

24. List three reasons it is important to properly handle and maintain hydraulic fluids.

____

____

____

25. The frequency of hydraulic system fluid quality inspections should be determined by the type of ______ and the intensity of ______.

____

26. The first step in testing a hydraulic fluid is to compare a sample taken from the reservoir with a sample of unused fluid of the same type and grade using a(n) ______ inspection.

____

27. Excessive oil temperatures in a hydraulic system using petroleum-based fluids will result in increased fluid ______ and lead to general fluid deterioration.

____

28. What are the three fundamental areas of system design and operation that must be analyzed to determine the cause of higher-than-desired system reservoir temperatures?

____

____

____

29. If the areas listed in question 28 are not the cause of higher-than-desired system reservoir temperatures, what other three factors, related to reservoir design, must be checked?

____

____

____

30. Systems that cannot maintain appropriate fluid temperatures using ambient air temperatures and appropriate reservoir, pump, control valve, and actuator designs employ specialized ______ devices to control temperature.

____

Activity 6-1

Analysis of Hydraulic Fluid Data Specification Sheets

Name ______________________ **Date** ____________

This activity is designed to show how a hydraulic fluid producer presents technical information about the fluids it markets. Hydraulic fluid specification sheets available from producers are used as the basis for study.

Activity Specifications

Analyze the data sheets of three hydraulic fluids designated by your instructor. The fluids should include a typical petroleum-based fluid, synthetic fluid, and water-based emulsion. Record the fluid name, number of the bulletin, and manufacturer's website (if appropriate). Then, carefully study the data sheets and answer the following activity questions.

Analyzed Fluid	Producer/Trade Name	Bulletin #	Website
#1	Castrol	Brayco Micronic 756	
#2	Cat Pumps	Crank Case oil	www.catpumps.com
#3	Renewable lubricants	Bio-HVO2''	www.renewablelube.com

1. Classify the three fluids described in the specification sheets by the primary material used in the formulation of the fluid (petroleum, water, etc.).
   #1 petroleum base
   #2 petroleum base
   #3 senthetic

2. Describe the operating conditions suitable for each of the three fluids, as claimed by the producer in the specification sheets.
   #1 Temp -65°F to 275°F operates in really low temps
   #2 avoid High temps & product contamination
   #3 Low temp- improves oxidation Stability, Low toxicity

3. What information is provided for each of the three fluids concerning handling and disposal?
   #1 Keep containers closed, wash thouraly after use, dispose where prohibited
   #2 Keep containers closed, store in cool, dry areas, dispose where prohibited
   #3 Keep containers closed, wash thoroughly after use, dispose where prohibited

4. List the ISO grades shown on the specification sheets for each of the fluids. If the sheet is describing a fluid series, provide data for each of the products shown.
   #1 ISO-15
   #2 ISO-68
   #3 ISO-46 & 68

5. Which systems other than ISO are used on these sheets to rate the viscosity of fluids? Why are several systems used to rate fluid characteristics?

6. Compare the viscosity index numbers for the fluids described on the specification sheets. According to the rating, which fluid of the three groups presented in this activity changes the least in viscosity as operating temperatures change? What is the basis for your answer?

1 Castrol Micronic 756 367
2 Cat pumps 135
3 Bio-HVO2 212
(Castrol) higher index

7. List the API gravity for each of the fluids shown on the specification sheets. If the sheet is describing a fluid series, provide data for each of the products shown. Compare the ratings for the fluids and indicate what information these data provides.

#1 Castrol 30.1
#2 cat pump 29.5
#3 Bio-HVO2 22.3

8. Discuss the fire-resistance characteristics of the three fluid groups included in this activity. Which indicators are used to describe this factor? What emphasis is placed on the characteristic in the literature and data?

#1 ______
#2 ______
#3 ______

9. Identify and describe the standards or test procedures used to indicate the lubricating and anti-wear characteristics of each of the three fluid groups included in this activity. Which units of measure or other methods are used on the specification sheets of the fluids to indicate the results of these tests?

#1 ______
#2 ______
#3 ______

10. Which standardizing group established the majority of the tests that provide the data for the information listed on the specification sheets used in this activity? What component manufacturers are referenced as providing tests that have also been used in evaluating the fluid?

#1 ______
#2 ______
#3 ______

Chapter 7

Source of Hydraulic Power

Power Units and Pumps

The activities in this lab exercise are designed to emphasize several of the principles discussed in the chapter, such as pump design features, manufacturer specifications, and basic power unit operating concepts. The activities are designed to provide exposure to the information provided by typical pump manufacturers. In addition, laboratory procedures illustrate concepts that can be applied to the operation of most pumps and power units.

Caution: Check with your instructor to be certain you are working with the correct equipment. Also, follow all safety procedures for your laboratory as you complete the activities.

Key Terms

The following terms are used in this chapter. As you read the text, record the meaning and importance of each. Additionally, you may use other sources, such as manufacturer literature, an encyclopedia, or the Internet, to obtain more information.

application information ______

axial-piston pump ______

balanced-vane pump ______

bent-axis design ______

cam ring ______

cavitation ______

centrifugal pumps ______

coupler ______

crescent design ______

cylinder barrel ______

dual pump ______

entrained air ______

external-gear pumps ______

fixed-delivery pumps ______

fluid filters ______

fluid flow weight ______

gear pumps

general hydraulic horsepower

general specifications

gerotor design

helical gear

herringbone gear

impeller

inline design

installation drawings

internal-gear pumps

jet pump

lobe pump

mechanical efficiency

motion-converting mechanism

non-positive-displacement pumps

overall efficiency

performance data

pintle

piston pumps

piston shoes

positive-displacement pumps

power unit

pressure balancing

pressure control valve

prime mover

prime mover horsepower

propeller pump

pump

radial-piston pumps ______________________________

____

____

reaction ring ______________________________

____

____

reciprocating pumps ______________________________

____

____

reciprocating-plunger pumps ______________________________

____

____

relief valve ______________________________

____

____

reservoir ______________________________

____

____

revolving-cylinder design ______________________________

____

____

rotary pumps ______________________________

____

____

safety valve ______________________________

____

____

screw pump ______________________________

____

____

shoeplate ______________________________

____

____

spur gear ______________________________

____

____

stationary-cylinder design ______________________________

____

____

strainer ______________________________

____

____

swash plate ______________________________

____

____

unbalanced-vane pump ______________________________

____

____

valve plate ______________________________

____

____

vane pumps ______________________________

____

____

vapor pressure ______________________________

____

____

variable-delivery pumps ______________________________

____

____

venturi ______________________________

____

____

volumetric efficiency ______________________________

____

____

wear plate ______________________________

____

____

Chapter 7 Quiz

Name ______________________________ Date ______________ Score ________

Write the best answer to each of the following questions in the blanks provided.

1. The two most common prime movers used in a hydraulic system are a(n) ______ and a(n) ______.

2. The ______ is the component in the power unit that stores hydraulic fluid when it is not being circulated through the system.

3. List the primary parts of a basic pump.

4. During the pressurization phase of basic pump operation, the ______ valve of the pump is closed.

5. What determines the output flow rate of a hydraulic pump?

6. A hydraulic pump that produces a constant output for each revolution of operation under varying system pressure is classified as a(n) ______-displacement pump.

7. List six factors that must be considered when selecting a pump for use in a hydraulic system.

8. In a gear pump, how is the reduced pressure required in the inlet chamber produced?

9. Describe the design of a gerotor gear pump.

____

____

____

____

10. A series of variable-sized pumping chambers are formed by the pump housing, cam ring, vanes, and ______ in a vane pump.

____

11. The ______ pump is the oldest of the pump designs used in current hydraulic system circuits.

____

12. Name the three classifications of the type of pumps in question 11.

____

____

____

13. The pistons of the inline axial-piston pump are powered by an angular device called a(n) ______ plate.

____

14. In a bent-axis piston pump, what establishes the piston stroke of the pump?

____

____

____

____

15. The type of axial-piston pump that is considered more rugged in performance is the ______ design.

____

16. The ______-piston pump design uses three or more cylinders symmetrically located around the power-input shaft of the pump.

____

17. The lubricating system design of many ______ pumps allows the pumping of materials that cannot be contaminated by petroleum-based lubricants.

____

18. Disadvantages of the lobe pump design are a relatively low operating pressure range and a(n) ______ fluid discharge flow.

____

19. List the four pump designs that can incorporate variable-flow delivery mechanisms in their construction.

____

____

____

____

20. Zero flow output can be achieved in a variable-displacement piston pump by adjusting pumping mechanism parts until the swash plate and cylinder barrel form an angle of ______ degrees.

____

21. A major advantage of using a pressure-compensated pump in a hydraulic system is reduced _____ consumption.

____

22. List four factors that can be improved in pumps that incorporate pressure balancing features.

____

____

____

____

23. Name the four categories of data typically found in information sheets provided by pump manufacturers.

____

____

____

____

24. Information concerning the shape and size of a pump should be found in the _____ section of pump information sheets.

____

25. List the three factors that are used in the formula to calculate general hydraulic horsepower.

____

____

____

26. List the four factors that are used to calculate the pressure in feet of head portion of the general hydraulic horsepower formula.

____

____

____

____

27. What are the two factors that make it necessary for actual prime mover horsepower to be higher than that calculated by the general horsepower formula?

____

____

28. *All* energy losses in the pump are indicated by the _____ rating.

____

29. The pressure at which a fluid begins to vaporize is known as _____ pressure.

____

30. Name the three conditions produced by pump cavitation.

____

____

____

Activity 7-1

Analysis of Hydraulic Pump Information Sheets

Name ______________________________________ Date _____________

This activity familiarizes you with the methods a fluid component manufacturer uses to present technical information about the pumps or power units it produces. The information is typically included in data sheets available in printed or electronic forms. The primary purpose of this activity is to familiarize you with the specifications of a pump series, rather than any of the additional components involved in a typical power unit.

Activity Specifications

Analyze pump or power unit data sheets designated by your instructor. Carefully study the data sheets and then answer the following activity questions based on the information presented in the sheets.

1. Describe the general characteristics of the pump or pump series, including the type of pumping mechanism, housing material, bearing type, and so on.

2. What is the continuous pressure rating of the pump series?

3. What is the operating speed rating of the pump series? In what type of installation would it be critical to know the range of acceptable operating speed?

4. Discuss the information provided relating to inlet line pressures. Why is there a difference in acceptable pressure drops between the indicated operating speeds? Why do you think the pressure is limited when the inlet line is charged?

5. Identify the following information about the pump. If there are several sizes of pumps in the series designated, indicate the specific model number on which you reported.
   A. Cubic inch displacement per revolution.

   ____

   B. Output flow at 1800 rpm and a system operating pressure of 2000 psi. If this specific information is not available, record the flow using similar data presented on the sheet. Be sure to note the pressure and speed settings.

   ____

6. Analyze the flow output of the *largest* of the pumps included in the pump series and answer the following questions.
   A. What happens to pump output as system pressure increases? Why?

   ____

   ____

   B. Compare the output at three different pump operating speeds. What happens to the effectiveness of the pump as the speed increases? Why?

   ____

   ____

7. Analyze the flow output of the *smallest* of the pumps included in the pump series and answer the following questions.
   A. Compare the performance of this pump model to the pump examined in question 6.

   ____

   ____

   ____

   B. Discuss any variations in data that indicate performance differences between the pump sizes.

   ____

   ____

   ____

8. What hydraulic fluid viscosity and temperature were used in preparing the performance data shown on this information sheet? Give specific examples of what changes will occur in the data if the viscosity and/or temperature is varied.

   ____

   ____

   ____

   ____

9. Select one pump model from the series described in the data sheets. Using appropriate performance information from the tables and graphs in the data sheet, estimate the input horsepower required to turn the pump at 2000 rpm with a system pressure of 2500 psi. Show your work.

   ____

10. Select another pump model from the data sheets. Using appropriate performance information from tables and graphs, estimate the output flow in gpm for the pump at 2500 rpm and a system pressure of 2000 psi. Show your work.

____

Activity 7-2

Pump Cavitation and Air Entrainment at the Inlet Line

Name ______________________________ **Date** ______________

Cavitation and air entrainment at the inlet line are problems that may be encountered during the operation of a hydraulic system power unit. Both of these conditions are undesirable and can cause damage to the pump if not identified and corrected. In this activity, you will create both conditions in a basic hydraulic power unit. This experience familiarizes you with the noise produced by the conditions and the testing that can be done to identify which condition is causing the problem.

Activity Specifications

Construct, operate, and record observations of the test circuit shown in the following diagram. This circuit is designed to illustrate pump cavitation and air entrainment at the inlet line. Check with your laboratory instructor before beginning this activity. The test circuit may already exist on selected test benches or power units in the shop.

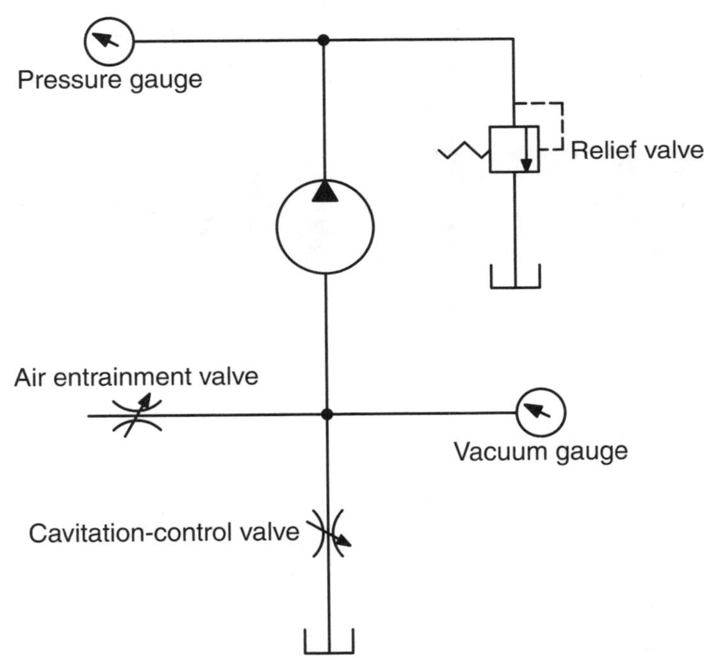

Required Components

Access to the following components is needed to complete this activity.

- Hydraulic power unit (pump, reservoir, and prime mover).
- Pressure control valve (relief valve).
- Pressure gauge.
- Vacuum gauge.
- Cavitation-control valve (flow control valve).
- Air entrainment valve (flow control valve).
- Manifolds (adequate number to make inlet line connections).
- Hoses (adequate number to allow assembly of the system).

Procedures

Study the following list of procedures to become familiar with the steps needed to complete the activity. Then, complete the procedure and record the observations indicated in the Data and Observations Record section.

1. Select the components needed to assemble the system and test circuit.
2. Assemble the system and have your instructor check the setup before proceeding.

Caution: Be certain the cavitation control valve is fully open and the air entrainment valve is fully closed before proceeding.

3. Start the power unit and adjust the relief valve of the circuit until the pressure gauge reads 500 psi.

Cavitation test and observation

4. Note and record the pressure reading on the inlet line vacuum gauge and the operating sound level of the pump. These observations can be considered the normal performance characteristics of the pump installation for the cavitation test.
5. Slowly close the cavitation control valve until the vacuum gauge reads 2″ Hg lower than the previous reading. Record this pressure reading and any changes in the operating sound of the pump.
6. Repeat step 5 increasing the vacuum gauge reading by 2″ Hg until the pump begins to cavitate. Cavitation is indicated by a substantial increase in noise from the pump.

Caution: Operate the power unit for only a short period (a few seconds) after cavitation begins as pump damage may result.

7. Note and record the specific inlet line vacuum gauge reading when pump noise indicates cavitation is occurring.
8. *Immediately* stop the power unit once cavitation is indicated.

Air entrainment test and observation

9. Completely open the cavitation-control valve.
10. Start the power unit. Observe and record the pressure reading on the inlet line vacuum gauge.
11. Slowly close the cavitation valve until the inlet line vacuum gauge reads 2 psi lower than the pressure observed in step 10. Note and record the operating sound level of the pump. These pressure and sound level observations can be considered the normal performance characteristics of the pump installation for the air entrainment test.
12. Open the air entrainment valve by 1/8 of a turn. Note and record the inlet line vacuum gauge pressure and the operating sound level of the pump.
13. Repeat step 12 noting the vacuum gauge reading and sound level for each adjustment.

Caution: Proceed carefully in opening the air entrainment valve to prevent a lose of pump prime, which can cause pump damage.

14. Stop the power unit as soon as a higher noise level is evident.
15. Discuss your data and observations with your instructor before disassembling the system.
16. Disassemble, clean, and return components to their assigned storage locations.

Data and Observation Records

Record pressure readings and general observations concerning pump cavitation and inlet line air entrainment.

Cavitation test record

A. Initial inlet line pressure: ______________________________

B. Test readings and observations

Test Setting	Inlet Line Pressure	Description of Pump Sound Level
1		
2		
3		
4		
5		
6		

C. Inlet line pressure at first indication of cavitation: ______________________

Air entrainment test record

A. Initial inlet line pressure: ____________________

B. Test readings and observations

Air Entrainment Valve Opening	Inlet Line Pressure	Description of Pump Sound Level
Closed		
1/8 turn open		
1/4 turn open		
3/8 turn open		
1/2 turn open		
5/8 turn open		
3/4 turn open		

C. Entrainment valve opening when pump noise initially increases: ____________________

D. Inlet line pressure when pump noise initially increases: ____________________

Activity Analysis

Answer the following questions in relation to the collected data and observation of the test operations.

1. Why is the initial inlet line pressure below atmospheric pressure in both the cavitation and air entrainment tests?

2. Explain the reason the pressure drops in the pump inlet line as the cavitation test valve is closed.

3. Why must the inlet line pressure be several psi below normal operating pressure before cavitation begins?

4. What causes the noise when a pump cavitates?

5. Describe what can mechanically happen to a pump if it continues to be operated while cavitating.

6. What happens to the inlet line pressure as the air entrainment valve is opened? Explain why this occurs.

7. What causes noisy pump operation when air enters the inlet line?

8. List several factors that could cause pump cavitation.

9. List several factors that could cause air entrainment at the pump inlet line.

10. What procedures would you use to identify the cause of noisy pump operation that may be caused by cavitation or air entrainment at the inlet line?

Chapter 8

Fluid Storage and Distribution

Reservoirs, Conductors, and Connectors

The activities in this lab exercise are designed to show several applications of the principles discussed in the chapter. The activities are based on the study of fluid storage and distribution components that relate to the hydraulic fluid power area. The activities provide exposure to a number of component design and construction factors, including the reservoir, conductors, and fittings.

Caution: Check with your instructor to be certain you are working with the correct equipment. Also, follow all safety procedures for your laboratory as you complete the activities.

Key Terms

The following terms are used in this chapter. As you read the text, record the meaning and importance of each. Additionally, you may use other sources, such as manufacturer literature, an encyclopedia, or the Internet, to obtain more information.

adapter fittings ______________________

baffle ______________________

braided wire ______________________

burst pressure ______________________

compression fittings ______________________

conductors ______________________

drain return line ______________________

Dryseal standard pipe thread ______________________

ferrule ______________________

fittings ______________________

flared fittings ______________________

fluid turbulence ______________________

hose ______________________

hose-end fitting ______________________

inlet line ______________________

laminar flow ______________________

manifolds ______________________________

____

____

nominal sizing ______________________________

____

____

packed slip joint ______________________________

____

____

pipe ______________________________

____

____

quick-disconnect coupling ______________________________

____

____

reservoir ______________________________

____

____

return line ______________________________

____

____

safety factor ______________________________

____

____

schedule number ______________________________

____

____

shock pressures ______________________________

____

____

spiral clearance ______________________________

____

____

spiral-wound wire tubing ______________________________

____

____

vortex ______________________________

____

____

Chapter 8 Quiz

Name ______________________________ Date ______________ Score ________

Write the best answer to each of the following questions in the blanks provided.

1. Name the five functions the reservoir performs in a hydraulic system.

____

____

____

____

____

2. A(n) ______ is used on reservoirs to allow the tank to take in or release air as the fluid level changes during system operation.

____

3. The bottom end of the return line in a reservoir is often cut at a 45° angle. Why is this done?

____

____

____

____

4. In general, the reservoir for a stationary industrial system should have a storage capacity ______ times the flow rate of the pump.

____

5. List three basic design factors that must be considered when selecting a fluid conductor for a hydraulic system.

____

____

____

6. Energy that is lost because of high flow resistance produces ______, which can cause system operating problems.

____

7. ______ pipe should not be used as a hydraulic system conductor as the coating on the pipe reacts with the additives in hydraulic oil.

____

8. The ______ sizing system states neither the actual inside nor outside diameter of a pipe.

____

9. The thin-walled, semirigid characteristics of tubing allow it to be bent to angles as sharp as ______. This allows fluid conductors with a minimum of connectors and fittings.

____

10. Flexible hydraulic hose is desirable in systems involving movable machine members or where severe ______ is a problem.

____

11. What are the three common types of materials used in the middle section of a flexible hydraulic hose to provide the strength required to withstand system pressure?

____

____

____

12. A(n) ______ is a machined or fabricated element that provides a common hydraulic fluid supply to several system components. It may consist of something as simple as a pipe with fittings or be a set of complex passageways machined into blocks of metal or other material.

____

13. Pipe threads taper ______ per foot to provide a positive seal at a threaded pipe connection.

____

14. Tube fittings are attached to system components by either tapered pipe threads or straight threads sealed by a flat compression washer or a(n) ______.

____

15. The most common angle used for flaring tubing in hydraulic systems is ______.

____

16. Describe four general methods used for securing an adaptor fitting when attaching hydraulic conductors to fittings and major system components.

____

____

____

____

17. The maximum operating pressure listed on the specification sheet of a conductor is typically ______ percent of the burst pressure.

____

18. Excessive fluid velocity in a working line may produce ______ flow, which can result in reduced system efficiency.

____

19. Name four types of technical information that may be found in most company catalogs related to hydraulic system conductors and fittings.

____

____

____

____

20. Name four factors that are important to consider when selecting hose for an installation.

____

____

____

____

Activity 8-1

Reservoir Construction and Sizing

Name ______________________________ **Date** ______________

This activity is designed to show how the principles discussed in the chapter related to the reservoir are applied in actual fluid power equipment. Any equipment that includes an easily accessible power unit, such as a hydraulic training bench, should provide the information needed for this activity. Check with your instructor to determine the hydraulic power unit that you should examine.

Activity Specifications

Measure the reservoir of the power unit to establish the height, length, and width. Determine the fluid level in the reservoir when the system is not operating and when it is in operation. Examine the pump to determine flow output, location in relation to the level of fluid in the tank, and type of connections used between the pump and reservoir. Note any other construction details related to the information discussed in this chapter, such as the location of the outlet and inlet lines, internal baffles, breathers, and other factors. Answer the following questions based on your observations and the data you collected.

1. Record the physical dimensions of the tank and the depth of the fluid.
   A. Reservoir dimensions
      Height: ______________________________
      Width: ______________________________
      Length: ______________________________
   B. Fluid depth
      System shut down: ______________________________
      System operating: ______________________________
2. Record the rated flow output of the pump in gpm:
   ____
3. Sketch the physical shape of the reservoir in the space below. Show the dimensions. Also, identify the locations of parts such as the filler, air breather, fluid level indicator, pump inlet line, return line, drain return line, and drain plug.

4. Which of the three general reservoir configurations does this unit best match? Why do you think this style of tank was used in the installation?

____

____

____

____

5. Calculate the capacity of the reservoir in gallons if the tank was allowed to completely fill. Show your work.

volume (in^3) = length (in) × width (in) × height (in)

$$\text{capacity (gal)} = \frac{\text{volume (in}^3\text{)}}{231 \text{ in}^3}$$

6. Calculate the number of gallons of fluid actually in the reservoir when the system is shut down and when it is operating. Does this amount of reservoir fluid fall within the general rule for establishing the size of a reservoir discussed in the chapter? Explain your answer.

Shut down gallons: _______________

Operating gallons: _______________

____

____

____

____

7. List at least three reasons that the normal hydraulic fluid level in the reservoir is well below the top of the tank.

____

____

____

____

____

____

8. Why is it recommended to have an air breather on the reservoir? What could happen to the pump if the air breather is omitted or becomes plugged during system operation?

____

____

____

____

____

____

9. Describe how the reservoir in this application helps to remove the heat generated by the system during operation. What special reservoir design features can be used to promote heat removal?

10. What design features are used with this reservoir to facilitate cleaning the inside of the reservoir when changing hydraulic system fluid?

Activity 8-2

Conductor Design, Construction, and Application

Name ______________________________ **Date** ______________

This activity is designed to show how the principles discussed in the chapter related to conductors are applied in actual fluid power equipment. The activity involves two sections. The first section involves an operating power unit to gain experience identifying hydraulic conductors in actual use. The second section involves information from conductor manufacturers that provides specific details on conductors.

Activity Specifications

The first portion of the activity involve the examination of the power unit of the fluid power training bench in your laboratory or another piece of operating hydraulic-power equipment assigned by your instructor. Identify the types and sizes of conductors used in the application (pipe, tubing, or flexible hose). Be certain to examine the pump to determine flow output. Note any other construction details that relate to the conductor information discussed in this chapter, such as fittings and ease of disassembly and assembly for service.

The second portion of the activity includes a section related to information available on hoses in manufacturer catalogs or Internet sites. This portion of the activity illustrates the type and amount of information available when selecting conductors for an installation.

Procedures

Your instructor will assign a hydraulic system power unit for part one of the activity and a website or catalog for part two. For each part of the activity, supply the requested information and answer the questions in relation to the assigned hydraulic equipment and manufacturer data.

Part one—system conductors

1. Record the type and sizes of the conductors used in the installation.

	Example 1		Example 2	
Type of Conductor	**Size**	**Application**	**Size**	**Application**
Pipe	________	________________	________	________________
Tubing	________	________________	________	________________
Flexible Hose	________	________________	________	________________
Manifold	________	________________	________	________________

2. Record the rated output of the pump in gallons per minute:
____

3. Calculate the fluid velocity in the pump inlet line using the information collected concerning conductor size and pump output. Show your work. Does the calculated velocity fall below the maximum velocity recommended for inlet lines specified in the chapter?

4. What problem could occur in the system if the calculated fluid velocity in problem 3 is higher than the recommended specification?

5. Calculate the fluid velocity in one of the working lines using the information collected concerning conductor size and pump output. Show your work. Does the calculated velocity fall below the maximum velocity recommended for working lines specified in the chapter?

6. What problem could occur in the system if the calculated fluid velocity in problem 5 is higher than the recommended specification?

7. What is the maximum fluid flow rate that can be transmitted while maintaining an acceptable fluid velocity for the working line used for the calculations in problem 5? Show your work.

____

8. In many systems, the flow velocity in working lines created by the pump output capacity is well below the maximum recommended velocity. Explain why this is a good design feature in a system.

____

____

____

____

Part two—conductor catalog or website descriptions

Most companies manufacture a large number of hoses. Check with your instructor to obtain three specific hose types for this portion of the activity. The ability to locate and interpret information from similar catalog and data sheets is a part of the skill set required by individuals working in the fluid power industry. Based on an actual catalog page or the information available on a manufacturer's website, provide the following information.

Designated hoses:

A. ____

B. ____

C. ____

1. What is the primary materials used in all three of these hose types to provide the strength to withstand system operating pressure?

____

____

2. What is the maximum operating pressure of each of these hose types when the inside diameter is 1/2″?

____

____

3. How is the maximum operating pressure affected as the inside diameter of the hose increases from 1/2″?

____

____

As it decreases from 1/2″?

____

____

4. Do any of the hose types indicate they can be used with fire-resistant hydraulic fluids? What is different in the construction of a hose that allows it to be used with a fire-resistant fluid?

5. Which of the three hose types has the greatest operating temperature range? What is that range?

6. As a general rule, what happens to the bend radius of a hose as its diameter increases? Use an example from any one of the data sheets to support your statement.

7. What is the safety factor between the recommended maximum operating pressure and the minimum burst pressure of these hoses?

Activity 8-3

Conductor Fitting Design and Application on Fluid Power Equipment

Name ______________________________ Date ______________

This activity is designed to show how the principles discussed in the chapter related to conductor fittings are applied in the construction of the fluid power distribution lines. The activity illustrates the wide variety of fitting types that are available for assembly of hydraulic fluid conductors. The activity involves two sections. The first section concerns an operating hydraulic unit to provide experience identifying hydraulic conductor fittings in actual use. The second section involves information provided by fitting manufacturers.

Activity Specifications

The first portion of the activity involves examining the power unit and manifolds of the fluid power training bench in your laboratory or another piece of operating hydraulic-powered equipment assigned by your instructor. Identify the types and sizes of conductor fittings used in the application. Be certain to examine each point where a pipe, tube, or flexible hose is joined to another conductor or component. Note any other construction details related to the conductor and conductor fitting information discussed in this chapter, such as the ease of disassembly and assembly for service.

The second portion of the activity is related to information available on fittings from manufacturer Internet sites. This portion of the activity illustrates the type and amount of information available when selecting fittings for pipe, tube, and hose installations.

Procedures

Your instructor will assign a hydraulic system power unit for part one of the activity and a website for part two. For each part of the activity, supply the requested information and answer the questions in relation to the assigned hydraulic equipment and manufacturer data.

Part one—conductor connections

1. Identify the type of conductor connections used in the installation. Note the general name of the fitting or adaptor and the sealing methods used in two of the connections.

	Fitting Example 1			Fitting Example 2		
Type of Conductor	**Name**	**Seal A**	**Seal B**	**Name**	**Seal A**	**Seal B**
Pipe						
Tubing						
Flexible Hose						
Manifold						

2. Which type of conductor was most frequently used in the hydraulic system you examined? Why do you think that particular type was selected for use on the equipment?

____

____

____

3. What is the principle involved that causes a pipe thread connection to seal?

4. Which style of pipe thread is recommended for use in hydraulic system connections? Why?

5. When straight threads are used on fittings and adaptors, what is used to provide a positive seal?

6. Which fitting is used in pipe to allow pipe sections to be disassembled without unscrewing the threaded connections? Describe the configuration of the sealing surfaces in this fitting.

7. What is a quick coupling and how does it serve as a fitting to connect conductors in a fluid power system?

Part two—catalog fitting and adapter descriptions

The following questions are based on two specific websites assigned by your instructor. One of these deals with hydraulic hose fittings and the other deals with hydraulic tube fittings. The websites should contain illustrations and charts providing detailed information on a number of fittings and adapters manufactured by the company. Numerous design variations are available from manufacturers, making service repairs and modifications somewhat challenging. The ability to recognize fitting types and interpret information presented in catalogs and data sheets is part of the skill set needed by individuals working in the fluid power industry.

Assigned website addresses:

Site 1, hose fittings: ____________________

Site 2, tube fittings: ____________________

1. Approximately how many different design variations of hose fittings are listed at site 1?

2. Approximately how many different design variations of tube fittings are listed at site 2?

3. Name and briefly describe two methods used to connect flexible hose to fittings that can be attached to other conductors and components. What are the advantages of each of these types of fitting methods?

   Method 1: ____________________

   Method 2: ____________________

4. List the various methods used by the manufacturer to provide a leakproof fit between metal tubing and hydraulic fittings.

5. Describe tube connectors that make use of flared surfaces for a seal. What type of equipment is needed to produce these flares?

6. Discuss the reasons for the number of various fittings (designs, sizes, materials) available to assemble fluid power circuits. Is this variety necessary and why or why not? Has it developed because of suppliers creating more-effective products? How is this related to internationalization of the manufacturing process?

Chapter 9

Actuators

Workhorses of the System

The activities in this lab exercise are designed to show several applications of the principles discussed in the chapter. The activities use hydraulic cylinders and motors to demonstrate these principles. Primary emphasis is placed on force and speed calculations for cylinders and the operation of motors in basic rotary actuator circuits.

Caution: Check with your instructor to be certain you are working with the correct equipment. Also, follow all safety procedures for your laboratory as you complete the activities.

Key Terms

The following terms are used in this chapter. As you read the text, record the meaning and importance of each. Additionally, you may use other sources, such as manufacturer literature, an encyclopedia, or the Internet, to obtain more information.

barrel ______________________________

cam-type radial-piston motors ______________________________

cap ______________________________

clevis mount ______________________________

closed-loop circuit ______________________________

cushioning ______________________________

cylinder ______________________________

double-acting cylinder ______________________________

double-rod-end cylinder ______________________________

dynamic seals ______________________________

effective area ______________________________

effective piston area ______________________________

fixed-centerline mount ______________________________

fixed-displacement motor ______________________________

fixed-noncenterline mount ______________________________

head ______________________________

hydraulic motor ______________________________

hydrostatic drive ______________________________

hydrostatic transmission ______________________________

inertia ______________________________

integral construction ______________________________

limited-rotation actuator ______________________________

linear actuator ______________________________

mill cylinders ______________________________

motor braking circuit ______________________________

nonintegral construction ______________________________

one-piece cylinder ______________________________

open-loop circuit ______________________________

orbiting gerotor motor ______________________________

parallel circuit ______________________________

pinion gear ______________________________

piston ______________________________

pivoting-centerline mount ______________________________

rack gear ______________________________

ram ______________________________

replenishment circuits ______________________________

rod ______________________________

rotary actuator ______________________________

screw motor ______________________________

series circuit ______________________________

side loading ______________________________

single-acting cylinder ______________________________

static seals ______________________________

stator ______________________________

telescoping cylinders ______________________

____

____

threaded-end cylinders ____________________

____

____

tie-rod cylinder __________________________

____

____

trunnion mount ___________________________

____

____

universal joint ___________________________

____

____

variable-displacement motor ________________

____

____

wear plates ______________________________

____

____

wiper seal _______________________________

____

____

Chapter 9 Quiz

Name ________________________ **Date** ____________ **Score** ________

Write the best answer to each of the following questions in the blanks provided.

1. Name the three basic components that make up a hydraulic cylinder.

2. The cylinder end through which the rod passes is called the ______, while the end without the rod is called the ______.

3. Name the two common ways in which cylinders are classified.

4. *True or False.* The volume of the blind end of a double-acting cylinder is smaller than the volume of the rod end because of the space taken by the rod.

5. List the four basic construction styles of cylinders.

6. Define *cushioning.*

7. Name the two pivoting-centerline cylinder-mounting methods that allow the cylinder to follow the arc made by a pivoting machine member.

8. What is the formula for calculating the force generated by a cylinder during extension?

9. Name three applications for limited-rotation actuators.

10. Which characteristic of a hydraulic motor indicates the volume of fluid required to turn the motor output shaft one revolution?

11. A motor in which the output shaft bearings are side loaded is classified as a(n) ______ design.

____

12. The design that is considered the simplest of the hydraulic motors is the ______ motor.

____

13. The specialized gear design that is used with most internal gear motors is called a(n) ______.

____

14. Name the design elements that are included in the housing of a fixed-displacement, pressure-unbalanced vane motor.

____

____

____

____

15. As the displacement is reduced in a variable-displacement vane motor, the speed of the motor increases, while the potential ______ output decreases.

____

16. How is the balancing feature achieved on the balanced-vane motor?

____

____

____

____

17. In the inline piston motor design, the centerline of the pistons and cylinders is inline with the centerline of the ______.

____

18. The centerlines of the pistons and cylinders and the power output shaft are at an angle in the ______ piston motor design.

____

19. A(n) ______ is used to connect the cylinder barrel and the power output shaft of a bent-axis hydraulic motor.

____

20. In the radial-piston motor with a rotating cylinder block, what acts as a bearing surface to support the rotation of the cylinder block and also houses the ports to control fluid entering and leaving the cylinders?

____

21. If hydraulic motors are connected in ______, all of them operate at the same time, as long as the total pressure requirement is below the setting of the system relief valve.

____

22. Stopping a hydraulic motor that is operating with a heavy, rotating load can be difficult because of the ______ of that load.

____

23. ______ circuits are used in closed-loop hydraulic systems to make up fluid lost from leakage during system operation.

____

24. Name four advantages of hydrostatic drives over conventional transmission systems.

____

____

____

____

25. Which hydrostatic transmission type typically combines the pump, motor, fluid lines, valving, and accessories into a single housing?

____

Activity 9-1

Hydraulic Cylinder Force and Speed

Name ______________________________ **Date** ______________

Cylinder force and operating speed are basic to the operation of most hydraulic circuits. Understanding the capability of a cylinder to generate force and speed is fundamental to selecting a cylinder that can efficiently operate within the requirements of the application. This activity requires the calculation of both force and speed. Then, a hydraulic test circuit is used to verify the theoretical calculations.

Activity Specifications

Calculate the theoretical force and speed that a cylinder is capable of generating during extension and retraction at a specific maximum system pressure and flow. Set up a test circuit to verify the results of the calculations and identify possible reasons for variations. The circuit diagram shown below includes the power unit and the necessary test components. The loading valve simulates a load on the cylinder by restricting flow out of the cylinder. It is used only during the force-measuring part of the activity.

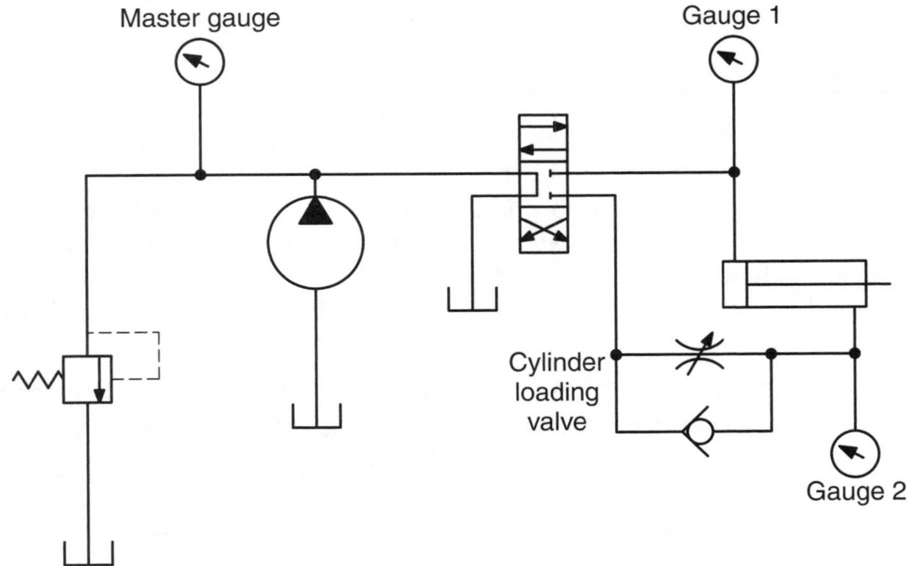

Required Components

Access to the following components is needed to complete this activity.

- Hydraulic power unit (pump, reservoir, and prime mover).
- Pressure control valve (relief valve).
- Directional control valve (four way, three position, tandem center).
- Flow control valve with return check valve (used as the cylinder loading device).
- Three pressure gauges.
- Cylinder (double acting, single rod).
- Manifolds (adequate number to make gauge and other connections).
- Hoses (adequate number to allow assembly of the system).

Procedures

Study the following procedures to become familiar with the steps needed to complete the activity. Then, complete the procedures and record the observations indicated in the Data and Observations Record section.

Component information/calibration

1. Select the components needed to assemble the system and test circuit.
2. Determine the bore, rod diameter, and stroke of the cylinder you have selected for use in the activity. Determine the rated output flow of the pump used in the hydraulic power unit. Record these dimensions and capacities in the Data and Observation Records section.

> **Note:** The dimensions and capacities may be shown on a label of the pump and cylinder or found in a manufacturer's catalog. In the case of the cylinder, it may be necessary to use a scale and micrometer to determine the specifications.

3. Assemble the system. Have your instructor check your selection of components and the setup of the circuit before proceeding.
4. Completely close the loading valve in the circuit.

> **Note:** The loading valve is used in this activity to restrict flow between the discharge side of the cylinder and the reservoir. Closing this valve simulates a load on the cylinder rod by creating back pressure in the cylinder chamber connected to the reservoir.

5. Start the power unit. While attempting to extend the cylinder, adjust the relief valve of the circuit until the master gauge reads 500 psi.
6. Open the loading valve. Check the calibration of gauges 1 and 2 by completely extending the cylinder. Hold the directional control valve in the shifted position and note the pressure on gauge 1. Completely retract the cylinder and note the pressure on gauge 2 when the directional control valve is held in the shifted position. Record these two pressures in the Data and Observation Records section.

> **Note:** If the two gauges are accurately calibrated, they should read very close to the same pressure. If there is a variation in the pressure readings, note the difference and correct the pressure used for your calculations.

Cylinder extension force observation and test

7. Fully retract the cylinder and completely open the loading valve.
8. Shift the directional control valve to extend the cylinder while adjusting the loading valve until gauge 1 reads 300 psi during cylinder extension. This can be accomplished by experimentation while shifting the directional control valve to extend and retract the cylinder several times. Once the loading valve has been set, observe and record the pressure on gauge 2 during extension.
9. Repeat step 8 to obtain a reading of 400 psi on gauge 1 during cylinder extension. Again note and record the pressure on gauge 2 during extension.

Cylinder retraction force observation and test

10. Change the position of the cylinder in the circuit so the rod end of the cylinder is connected to gauge 1 and the blind end is connected to gauge 2.

> **Note:** Shifting the cylinder connections allows the retraction phase of cylinder operation to be tested with a minimum of circuit changes.

11. Fully extend the cylinder and completely close the loading valve.
12. Shift the directional control valve to retract the cylinder while adjusting the loading valve until gauge 1 reads 300 psi during cylinder retraction. Once the loading valve has been set, note and record the pressure on gauge 2 during retraction.
13. Repeat step 12 to obtain a reading of 400 psi on gauge 1 during cylinder retraction. Again note and record the pressure at gauge 2 during retraction.

Cylinder extension and retraction speed

14. Remove gauges 1 and 2 and the loading valve from the circuit and reconnect the cylinder.
15. Completely retract the cylinder.
16. Shift the directional control valve to extend the cylinder. Carefully note the time the cylinder takes to extend and the maximum pressure on the master gauge while the cylinder rod is extending. Record this time in the Data and Observation Records section.
17. Shift the direction control valve to retract the cylinder. Carefully note the time and pressure required for retraction. Record this time in the Data and Observation Records section.
18. Repeat steps 15 through 17 to check the accuracy of the initial data collected.
19. Discuss your data and observations with your instructor before disassembling the system.
20. Disassemble, clean, and return components to their assigned storage locations.

Data and Observation Records

Record component information, pressure readings, times, and general observations concerning cylinder force and speed.

Component information/calibration record

A. Cylinder bore: ____________________

B. Rod diameter: ____________________

C. Cylinder stroke: ____________________

D. Gauge pressure calibration:

Gauge 1: ____________________

Gauge 2: ____________________

E. Pressure correction:

Gauge 1: ____________________

Gauge 2: ____________________

F. Pump flow rate: ____________________

Cylinder extension force test record

Test Number	Pressure Readings	
	Gauge 1	Gauge 2
1	300 psi	
2	400 psi	

Cylinder retraction force test record

Test Number	Pressure Readings	
	Gauge 1	Gauge 2
1	300 psi	
2	400 psi	

Cylinder extension and retraction speed test record

Test	Cylinder Rod Travel Time	Master Gauge Pressure
Extension 1		
Extension 2		
Retraction 1		
Retraction 2		

Activity Analysis

Answer the following questions in relation to the data collected and observation of the test procedures.

1. Using the cylinder specifications identified during this activity, calculate the theoretical force that can be produced during extension and retraction at the specified system relief valve setting. Show your work.

   Extension: ______________________________

   Retraction: ______________________________

2. Examine the results of the calculations in question 1. Does the maximum cylinder force vary between the extension and retraction strokes? Why or why not? ____________________

____

____

____

____

3. Calculate the force generated by the test cylinder during *extension* when gauge 1 pressure is 300 psi. Based on the gauge 2 reading, calculate the resisting force generated by the loading valve back pressure in the rod end of the cylinder. Repeat these calculations for the second test completed with gauge 1 reading a pressure of 400 psi. Show your work. Be certain to adjust for any variations in the pressure gauge readings identified in the first part of this activity.

Cylinder extension force at 300 psi: ____________________

Loading valve back pressure at 300 psi: ____________________

Cylinder extension force at 400 psi: ____________________

Loading valve back pressure at 400 psi: ____________________

4. Calculate the force generated by the test cylinder during *retraction* when gauge 1 pressure is 300 psi. Based on the gauge 2 reading, calculate the resisting force generated by the loading valve back pressure in the blind end of the cylinder. Repeat these calculations for the second test completed with gauge 1 reading a pressure of 400 psi. Show your work. Be certain to adjust for any variations in the pressure gauge readings identified in the first part of this activity.

   Cylinder retraction force at 300 psi: ______________________

   Loading valve back pressure at 300 psi: ______________________

   Cylinder retraction force at 400 psi: ______________________

   Loading valve back pressure at 400 psi: ______________________

5. Examine the results of the calculations in questions 3 and 4. Compare the resistance force generated by the loading valve back pressure and the force that is moving the cylinder piston and rod. Should these figures be the same or different? Justify your answer.

   ____

   ____

   ____

   ____

6. Using the cylinder specifications identified during this activity, calculate the theoretical cylinder extension and retraction speeds that can be achieved with the test cylinder considering the pump flow rate identified. Show your work.

   Extension speed: ______________________________

   Retraction speed: ______________________________

7. Convert the cylinder rod travel times that you measured in this activity into cylinder extension and retraction speeds. Show your work. Compare these measured speeds with the calculated speeds. List several factors that may cause variations between the calculated and the actual speeds encountered.

   ____
   ____
   ____
   ____

8. During the measurement of cylinder speed, the pressure was noted on the master gauge. How does system pressure fit into the overall picture when setting up a circuit where careful control of cylinder speed is critical to system operation?

   ____
   ____
   ____
   ____

Activity 9-2

Hydraulic Motor Series and Parallel Circuits

Name ______________________________ Date ______________

Multiple hydraulic motors may be connected in a circuit in a variety of ways to produce desired system operation. Series and parallel connection of motors are basic concepts that produce very different operating characteristics. An understanding of how these basic circuits perform under varying loads is essential to a full understanding of actuator operation in a hydraulic system.

Activity Specifications

This activity involves the identification of the specifications of a hydraulic motor and the construction and operation of two hydraulic motor test circuits. The first circuit uses two hydraulic motors connected in series. The second circuit uses two hydraulic motors connected in parallel. A simulated load is placed on the motors in each circuit and observations are made concerning motor speed, fluid flow, and system pressure. The motors are placed under load, tested to determine system pressure, and their speed determined. Flow control valves are used as loading valves to simulate a load on the motor by restricting flow out of the motors.

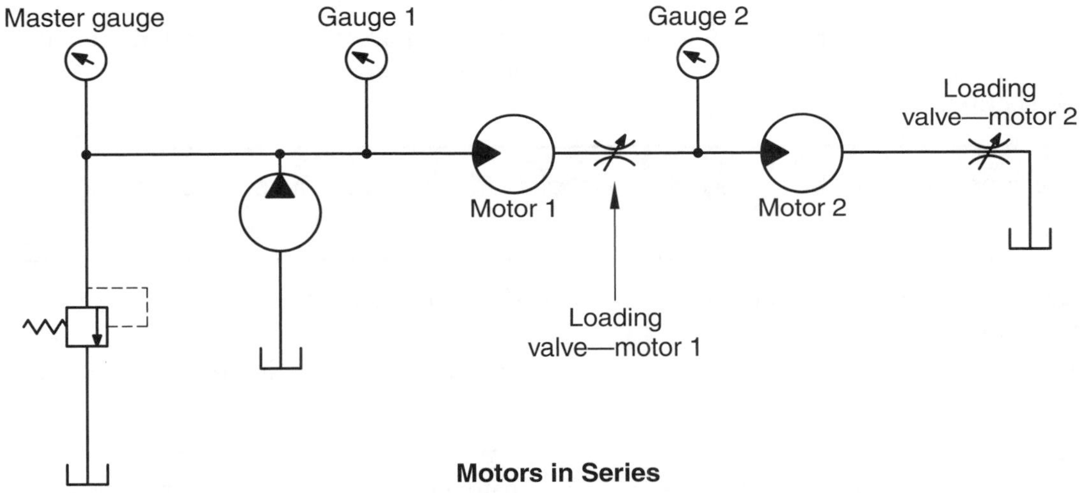

Motors in Series

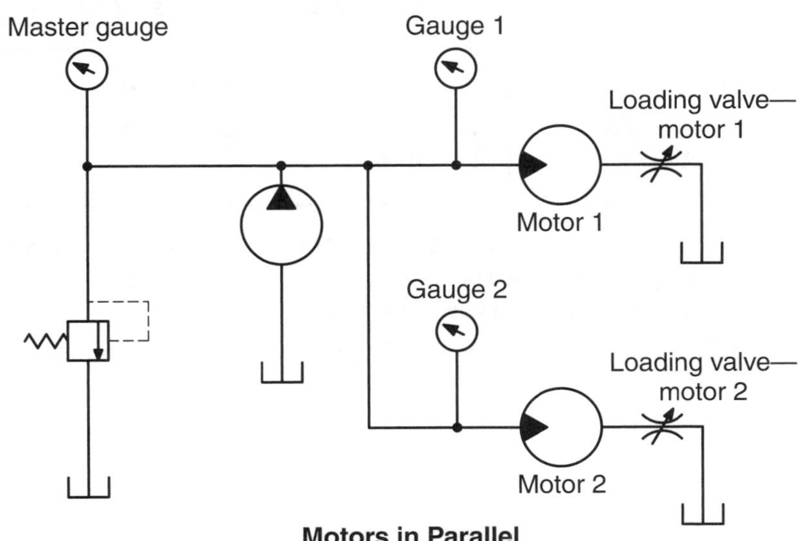

Motors in Parallel

Required Components

Access to the following components is needed to complete this activity.

- Hydraulic power unit (pump, reservoir, and prime mover).
- Pressure control valve (relief valve).
- Two flow control valves without return check valves (used as the motor loading devices).
- Three pressure gauges.
- Two hydraulic motors (if possible, use motors with the same displacement).
- Manifolds (adequate number to make gauge connections).
- Hoses (adequate number to allow assembly of the system).
- Tachometer (hand-held, reflective-tape type; used to measure output shaft speed).

Procedures

Study the following procedures to become familiar with the steps needed to complete the activity. Then, complete the procedures and record the observations indicated in the Data and Observations Record section.

Component information/calibration

1. Select the components needed to assemble the motor test circuits.
2. Identify the displacement of the motors you have selected. Determine the rated output flow of the pump used in the hydraulic power unit. Record these capacities in the Data and Observation Records section.

> **Note:** The capacities may be displayed on a label on the pump and motor or found in a manufacturer's catalog.

3. Assemble the series test circuit and have your instructor check your selection of components and the setup of the circuit before proceeding.
4. Completely *open* the loading valve for motor 1 in the circuit and completely *close* the loading valve for motor 2.

> **Note:** The loading valves are used in this activity to restrict flow between the discharge side of the motors and the reservoir. Closing the valve simulates a load that will stall the motor.

5. Start the power unit and adjust the relief valve of the circuit until the master gauge reads 500 psi.
6. Check the calibration of gauges 1 and 2. Record the pressures of both gauges in the Data and Observation Records section. Stop the power unit.

> **Note:** If the two gauges are accurately calibrated, they should read the same pressure. Under this test condition, any variation in the pressure readings indicates a variation in the calibration of the gauges. Note the difference and correct the pressure used for your calculations.

Circuit one—series connection

7. Remove any load from the motors by completely opening both loading valves. Start the power unit.
8. Observe and record the pressure on the master gauge, gauge 1, and gauge 2. Measure and record the speed of each motor.
9. Load motor 1 by slowly closing the loading valve for the motor until gauge 1 reads 250 psi. Note and record the pressure on each of the gauges and the speed of each motor.
10. Increase the load on motor 1 by slowly closing its loading valve until the motor stalls. Note the variations in system pressures and motor speed as the load increases on the motor. Record the gauge pressures and the speed of each motor after motor 1 stalls.
11. Decrease the load on motor 1 by opening its loading valve until gauge 1 reads 250 psi. Increase the load on motor 2 by slowly closing its loading valve until gauge 2 reads 150 psi. Note and record the pressure on each of the gauges and the speed of each motor.
12. Remove the load from motor 1 by completely opening the loading valve for the motor. Note and record the pressure at each of the gauges and the speed of each motor.
13. Slowly increase the load on motor 2 by closing the loading valve for the motor until it is completely shut. Note the variations in system pressures and motor speed as the valve is closed. Record the gauge pressures and the speed of the motors when the loading valve is completely shut. Stop the power unit.

Circuit two—parallel connection

14. Reconfigure the circuit as shown in the parallel test circuit. Adjust the relief valve for a maximum system operating pressure of 500 psi. Have your instructor check your selection of components and the setup of the circuit before proceeding.
15. Remove any load from the motors by completely opening both loading valves. Start the power unit.
16. Observe and record the pressure on the master gauge, gauge 1, and gauge 2. Measure and record the speed of each motor.
17. Increase the load on motor 1 by slowly turning its loading valve until the motor stalls. Observe the speed of the motors and changes in system pressure as the load increases. Note and record the pressure on each of the gauges and the speed of each motor when motor 1 is fully loaded.
18. Remove the load from motor 1 by completely opening the loading valve for the motor.
19. Increase the load on motor 2 by slowly closing its loading valve until the motor stalls. Observe the speed of the motors and changes in system pressure as the load increases. Note and record the pressure on each of the gauges and the speed of each motor when motor 2 is fully loaded.
20. Remove the load from motor 2 by completely opening the loading valve for the motor.
21. Balance the loads on the two motors by slowly increasing the load on each motor. To do so, equally turn the loading valves in small increments. Adjust the loads on the motors so gauge 1 and gauge 2 read 300 psi and the speed of each motor is equal. Record the pressure in the system and the speed of the motors in the Data and Observation Records section.
22. Slightly vary the load on motor 1. Adjust its loading valve by 1/8th turn increments *without* changing the setting of the load on motor 2. Note and record the pressure in the system and the speed of the motors as the load is varied.
23. Stop the power unit. Discuss your data and observations with your instructor before disassembling the system.
24. Disassemble, clean, and return components to their assigned storage locations.

Data and Observation Records

Record component information, pressure readings, times, and general observations concerning motor circuit operation.

Component information/calibration record

A. Motor displacement

Motor 1: ______________________________

Motor 2: ______________________________

B. Pump flow rate: ______________________________

C. Gauge pressure calibration:

Gauge 1: ______________________________

Gauge 2: ______________________________

D. Pressure correction:

Gauge 1: ______________________________

Gauge 2: ______________________________

Series motor connection test record

Activity Step	Motor Load		Gauge Pressure Readings			Motor RPM	
	One	Two	Master	One	Two	One	Two
8	no load	no load					
9	partial load	no load					
10	full load	no load					
11	partial load	partial load					
12	no load	partial load					
13	no load	full load					

Parallel motor connection test record

Activity Step	Motor Load		Gauge Pressure Readings			Motor RPM	
	One	Two	Master	One	Two	One	Two
16	no load	no load					
17	full load	no load					
19	no load	full load					
21	equal	equal					
22	1/8 turn increase	unchanged					
	1/4 turn increase	unchanged					
	3/8 turn increase	unchanged					

Activity Analysis

Answer the following questions in relation to the data collected and observation of the test procedures.

1. Using the pump and motor specifications identified during this activity, calculate the maximum speed this motor/pump combination can produce. Show your work.

   Speed (rpm): ______________________________

2. Check the calculated speed against the measured maximum speed obtained during operation of the test circuit. List several reasons why the calculated and actual operating speed may vary.

   ____
   ____
   ____
   ____

3. Explain why motors connected in series continue to rotate at identical speeds even when only one motor is loaded.

   ____
   ____
   ____
   ____

4. In the series circuit, why does the pressure at gauge 2 drop to 0 psi when motor 1 stalls?

   ____
   ____
   ____
   ____

5. In the series circuit, why does the pressure on gauge 1 increase to maximum system pressure when motor 2 is loaded until it stalls?

   ____
   ____
   ____
   ____

6. Describe what happens in the parallel circuit when both motors are operated without loads. Why do you think the motors operate in this way? Describe one other type of condition that could have happened.

   ____
   ____
   ____
   ____

7. What happens in the parallel circuit as the load on either of the motors is increased? Why?

8. Examine the pressure readings of the various tests done on the two circuits. Describe any patterns that explain the differences between these two circuit designs.

9. Which circuit design would produce the highest motor speed (rpm) with the power unit and motors selected for this activity? Why?

10. Describe at least one application suitable for each of these circuits. Justify your selection of these applications using operating factors demonstrated in this activity.

Controlling the System

Pressure, Direction, and Flow

The activities in this lab exercise are designed to show several applications of the control valve principles discussed in the chapter. The activities demonstrate principles relating to pressure, direction, and flow control valves. They apply the principles to both component and circuit design. Emphasis is placed on demonstrating those concepts key to understanding the effective operation of a full range of component, circuit, and system designs.

Caution: Check with your instructor to be certain you are working with the correct equipment. Also, follow all safety procedures for your laboratory as you complete the activities.

Key Terms

The following terms are used in this chapter. As you read the text, record the meaning and importance of each. Additionally, you may use other sources, such as manufacturer literature, an encyclopedia, or the Internet, to obtain more information.

balancing-piston valve ____________________

ball valve ____________________

brake valve ____________________

bypass flow control valve ____________________

check valve ____________________

compound relief valve ____________________

control elements ____________________

control valve ____________________

counterbalance valve ____________________

cracking pressure ____________________

detents ____________________

directional control valve ____________________

direct-operated valve ____________________

external force ____________________

fixed orifice ____________________

flow control valve ____________________

four-way valve

full-flow pressure

gate valve

globe valve

hydraulic pressure fuse

internal force

lands

needle valve

normal operating position

normally closed

normally open

orifice

pilot pressure

poppet valve

pressure compensation

pressure control valve

pressure override

pressure-reducing valve

pump unloading control

relief valve

restriction check valve

restrictor flow control valve

safety valves

sequence valve

shut-off valve

spool

spool valve

temperature compensation

three-way valve ______________________________
____
____

two-position, four-way valve ________________
____
____

valve body ___________________________________
____
____

Chapter 10 Quiz

Name ______________________ Date ____________ Score ________

Write the best answer to each of the following questions in the blanks provided.

1. Name the three control factors produced by the various types of valves in a hydraulic circuit.

2. Controlling the flow rate of fluid entering an actuator controls the ______ of that actuator.

3. In hydraulic components, a spool or piston often operates in a precision-machined ______ without using seals to prevent leakage between the parts.

4. List the three sources of external force that can be used to position control valve elements.

5. If the outlet line of a control valve is subjected to system operating pressure, it must use a(n) ______ drain.

6. Name the three basic uses of springs in the operation of hydraulic control valves.

7. The pressure at which a pressure control valve just opens is called the ______ pressure of the valve.

8. Heavy internal springs are used to establish the operating pressure of the ______-operated type of pressure control valve.

9. A compound relief valve commonly makes use of a(n) ______ and a(n) ______ in its operation.

10. Describe a *hydraulic pressure fuse.*

11. A(n) ______ drain must be used with sequence valves since both the initial and secondary sides of the valve are subjected to system pressure at the same time during system operation.

12. Counterbalance and brake valves provide ______ control in hydraulic systems.

____

13. Brake valves are used in some hydraulic motor circuits to prevent ______ loads from turning the motor at a higher-than-desired speed.

____

14. An unloading relief valve uses a(n) ______ mechanism to assist in the operation of the balancing piston in the pilot section of the valve.

____

15. A(n) ______ valve restricts a portion of a hydraulic circuit to a maximum operating pressure that is below the setting of the system relief valve.

____

16. What are the four general classifications of directional control valves used in a hydraulic system?

____

____

____

____

17. List the five common types of shut-off valves used in a hydraulic system.

____

____

____

____

____

18. ______ valves are used in hydraulic systems as both shut-off and metering devices.

____

19. Restriction check valves allow restricted flow through a line in one direction while allowing ______ flow in the reverse direction.

____

20. ______ directional control valves are designed for the actuation of rams and single-acting cylinders.

____

21. The ______ directional control valve is generally used in systems with double-acting cylinders or hydraulic motors.

____

22. Which type of the valve named in question 21 allows cylinder extension in one position and cylinder retraction in the second position?

____

23. There are five commonly available center positions for a three-position, four-way directional control valve. List these.

____

____

____

____

____

24. What are the five general categories of actuation methods used to shift directional control valves?

25. Describe a *solenoid*.

26. What is often used on the ends of the spool in directional control valves to automatically return the valve to the normal or restart position whenever the activation force is removed?

27. List the three factors that control the rate of fluid flow through an orifice.

28. Restrictor-type, noncompensated flow control valves provide a consistent flow rate as long as the load on the system is constant and the ______ of the fluid does not change.

29. An integral ______ valve is often used with needle valves to allow unrestricted reverse flow around the metering orifice.

30. What is a *biasing spring*, as related to a pressure compensator?

31. Name two common methods used in temperature-compensated flow control valves to maintain constant flow as the temperature of the system fluid varies.

32. Bypass flow control valves are often referred to as ______ because of their additional port.

33. Flow divider valves are flow control valves that divide one fluid supply between how many circuit branches or subsystems?

34. A(n) ______ flow divider valve allows pump output to be divided between system subsystems using a preselected ratio.

35. A system pump is delivering 10 gpm to the inlet port of a proportional flow divider valve. What amount of fluid is delivered to outlet port A and outlet port B if the delivery proportion is 50-50? What is the delivery if the proportion is 60-40?

Ratio	Port A	Port B
50-50	________	________
60-40	________	________

Activity 10-1

Analysis of Hydraulic System Valve Information Sheets

Name ______________________________ **Date** ______________

This activity is designed to show how a fluid power component manufacturer presents technical information about the valves it designs, manufactures, and markets. Most manufacturers supply considerable information about their valves, although the exact content and presentation vary between companies. This can even vary between models produced by the same organization.

Activity Specifications

Analyze the data sheets of three hydraulic valves designated by your instructor. The first of the valves should be a compound relief valve. The second valve should be a basic four-way, three-position directional control valve. The third valve should be a pressure-compensated flow control valve. The assigned valve models should have comparable flow capacities. Also, be certain the valves have similar pressure ratings. Carefully study the data sheets and then answer the activity questions.

Analyzed Valves

A. Relief valve

Manufacturer: ______________________________

Model: ______________________________

Pressure range: ______________________________

Flow capacity: ______________________________

B. Directional control valve

Manufacturer: ______________________________

Model: ______________________________

Pressure range: ______________________________

Flow capacity: ______________________________

C. Flow control valve

Manufacturer: ______________________________

Model: ______________________________

Pressure range: ______________________________

Flow capacity: ______________________________

Activity Analysis

Answer the following questions in relation to the collected data.

1. Describe the general construction characteristic of the valves. Compare information such as housing material, construction technique, and types of port connectors.

____

____

____

____

____

____

2. What options are available for mounting the valves in hydraulic circuits? Which valve type has the most available mounting options? Why?

3. List the types of seal materials available with these three valves. Why is it important to have a variety of these materials available when selecting valves?

4. What is the hydraulic oil viscosity range recommended for these valves? List the other oil properties recommended for these valves.

5. What options are available for adjusting the maximum operating pressure of the pressure relief valve? List and describe each control style that is available for the valve model selected for this activity.

6. Identify and list the operating pressure range options available for each of the three valve types. Which type valve has the most options available? Do you think this variation is typical between the three valve types studied in this activity? Why or why not?

7. Identify the methods that are available to shift or adjust the three valves studied in this activity. Divide the options into separate lists showing the methods available for relief, directional control, and flow control valves.

Relief

Directional control ____

____

____

Flow control ____

____

____

8. Identify the center position options available for the three-position directional control valve. Simply list the basic open, closed, and tandem center styles, if they are available. However, show the symbol and provide a brief description of the operation and possible application of any other available center design options.

____

____

____

____

____

____

9. Identify the rated accuracy of fluid-flow control through the pressure-compensated flow control valve when it is operated within rated operating conditions. Using data provided in the specification sheet, describe changes to flow control accuracy that may be expected if fluid pressure and temperature conditions vary during operation.

____

____

____

____

____

____

Activity 10-2

Pressure Control Valve Cracking and Pressure Override

Name ______________________________ **Date** ______________

This activity illustrates the concepts of cracking pressure, full-flow pressure, and pressure override for a pressure control valve. These terms are typically used to describe the operation of relief valves, although they also are basic to other normally closed pressure control valves, such as sequence and counterbalance valves. The activity provides experience in working with basic valve test procedures as well as the simple procedures necessary to plot the performance of a pressure control valve.

Activity Specifications

Set up a test circuit to measure the performance of hydraulic system relief valves. Plot and compare the performances of a simple, direct-operated and a more-complex, pilot-operated valve. Use the test circuit and procedure shown below to test the performance of the valves. Compare the test results to the specifications available from the valve manufacturer.

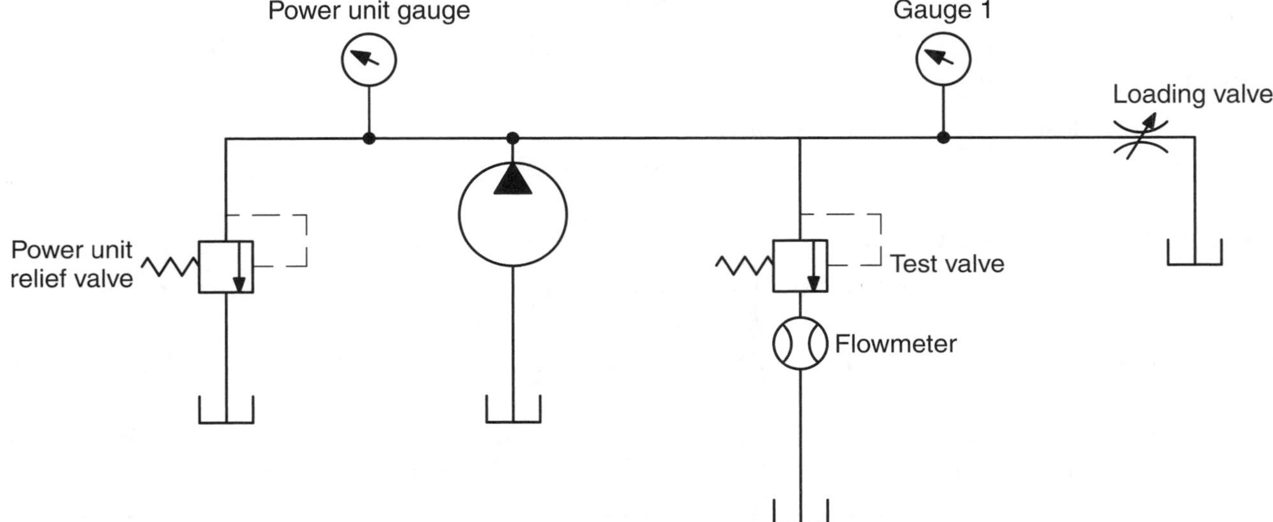

Required Components

Access to the following components is needed to complete this activity.

- Hydraulic power unit (pump, reservoir, and prime mover).
- Power unit relief valve.
- Two pressure gauges (one used as the power unit master pressure gauge and the second as the test pressure gauge).
- Flow control valve (used as the loading device for the test circuit).
- Flowmeter (used to identify the flow discharged to the reservoir through the test relief valves).
- Two relief valves (used as test valves in activity; one direct-operated valve and one pilot-operated valve).
- Manifolds (adequate number to make gauge connections).
- Hoses (adequate number to allow assembly of the system).

Procedures

Study the following procedures to become familiar with the steps needed to complete the activity. Then, complete the procedures and record the observations indicated in the Data and Observations Record section.

1. Select the components needed to assemble the system and test circuit. Collect and record catalog data on the relief valves.
2. Assemble the system using the direct-operated relief valve as the test valve. Have your instructor check your selection of components and the setup of the circuit before proceeding.
3. Disconnect the test valve and flowmeter from the circuit.
4. Adjust the relief valve of the power unit to the lowest pressure setting.
5. Completely open the loading valve of the test circuit.

Set the power unit relief valve maximum pressure

6. Start the power unit.
7. Slowly, completely close the test circuit loading valve.
8. Adjust the power unit relief valve so the power unit operates at 500 psi.

Set the test pressure for the direct-operated relief valve

9. Shut off the power unit and reconnect the test valve and flowmeter to the circuit.
10. Set the test valve at the lowest operating pressure.
11. Completely close the circuit loading valve.
12. Start the power unit and adjust the setting of the test relief valve until pressure gauge 1 reads 200 psi.

Test the performance of the direct-operated relief valve

13. Completely open the circuit load valve. Record gauge 1 pressure and the flow through the flowmeter in chart 1 of the Data and Observation Records section.
14. Slowly close the circuit load valve to increase the pressure on gauge 1 by 20 psi. In chart 1, record the pressure and flow through the flowmeter.
15. Proceed by closing the circuit load valve to increase the pressure in 20 psi increments. Record the pressure and flow readings at each setting until the 200 psi setting of the test valve is reached.

> **Note:** Record the exact pressure at which flow is first observed through the flowmeter. This is the *cracking pressure* of the valve.

16. Follow steps 9 through 12 to set the direct-operated relief valve to 300 psi.
17. Repeat steps 13 through 15 at the 300 psi setting. Record all pressure and flow readings in chart 2.

Test the performance of the pilot-operated relief valve

18. Shut off the power unit and replace the direct-operated relief valve in the circuit with a pilot-operated relief.
19. Set the test valve at the lowest operating pressure.
20. Completely close the circuit load valve.
21. Start the power unit and adjust the setting of the test relief valve until pressure gauge 1 reads 200 psi.
22. Completely open the circuit load valve. Record gauge 1 pressure and the flow through the flowmeter in chart 1 of the Data and Observation Records section.

23. Slowly close the circuit load valve to increase the pressure on gauge 1 by 20 psi. Record the pressure and flow through the flowmeter in chart 3.
24. Proceed by closing the circuit load valve to increase the pressure in 20 psi increments. Record the pressure and flow readings at each setting until the 200 psi setting of the test valve is reached.
25. Repeat the procedure in this section with the test relief valve set at 300 psi. Record gauge 1 pressure in chart 4.

Activity completion

26. Discuss your data and observations with your instructor before disassembling the test circuit.
27. Clean and return all component parts to assigned storage locations.
28. Complete the activity questions.

Data and Observation Records

Before conducting the laboratory test procedures, use manufacturer catalogs to collect information about your test relief valves. Record the basic information below. Then, carefully record the pressure and flow information from the test procedures in the charts provided.

Component information

	Direct-Operated Valve	Pilot-Operated Valve
Brand		
Model number		
Pressure rating		
Flow capacity		

Direct-operated valve test records

Chart 1 200 psi setting		Chart 2 300 psi setting	
Pressure (psi)	Flow (gpm)	Pressure (psi)	Flow (gpm)

Cracking pressure: ______________________________

Pilot-operated valve test records

Chart 3 200 psi setting		Chart 4 300 psi setting	
Pressure (psi)	Flow (gpm)	Pressure (psi)	Flow (gpm)

Cracking pressure: ______________________________

Activity Analysis

Complete the following in relation to the data collected and observations of the test procedures.

1. Plot the flow returned to the reservoir through the direct-operated relief valve at the 200 psi setting. Use the data you entered in chart 1.

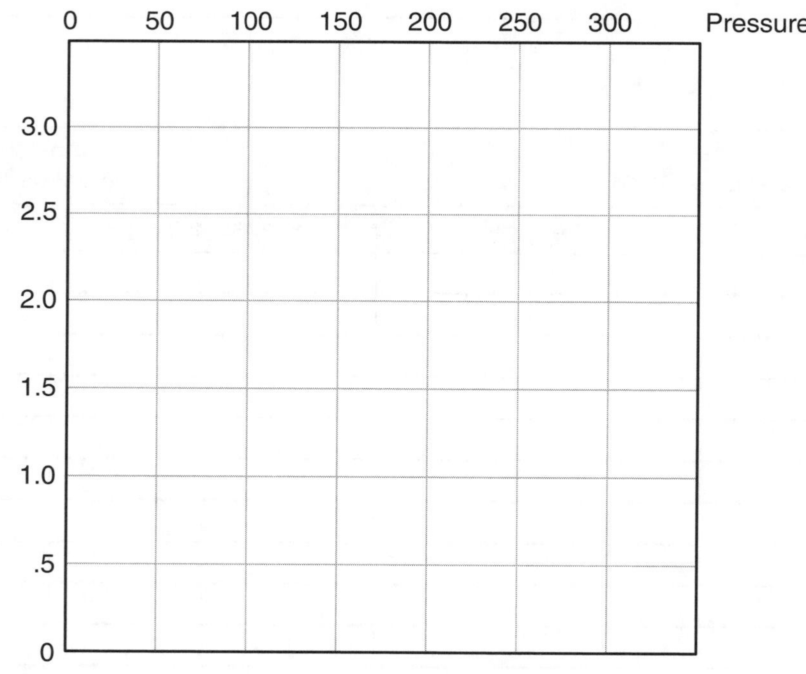

2. Plot the flow returned to the reservoir through the direct-operated relief valve at the 300 psi setting. Use the data you entered in chart 2.

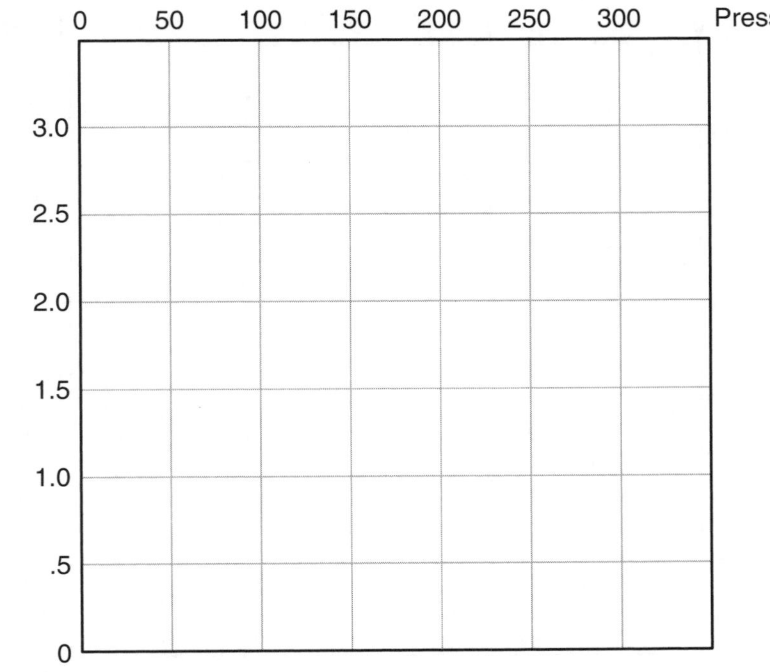

3. Plot the flow returned to the reservoir through the pilot-operated relief valve at the 200 psi setting. Use the data you entered in chart 3.

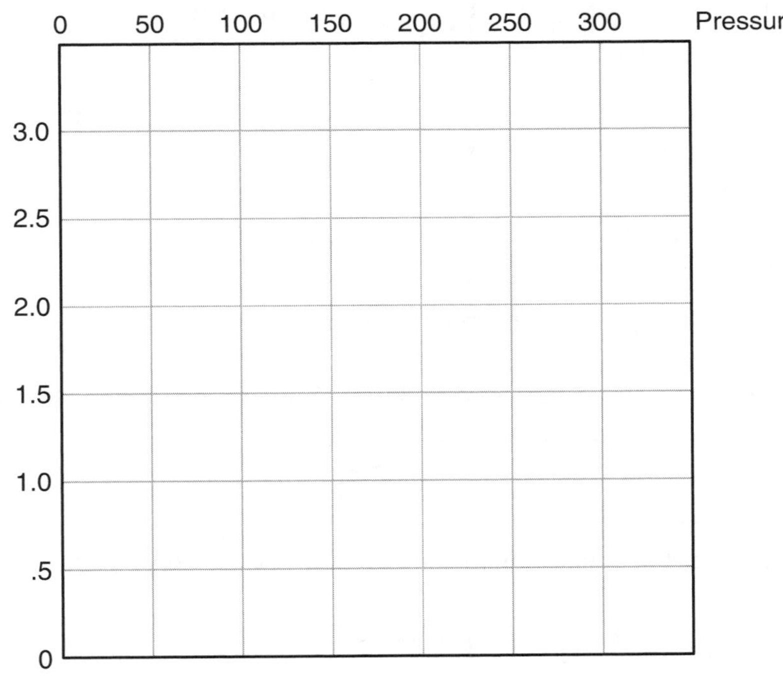

4. Plot the flow returned to the reservoir through the pilot-operated relief valve at the 300 psi setting. Use the data you entered in chart 4.

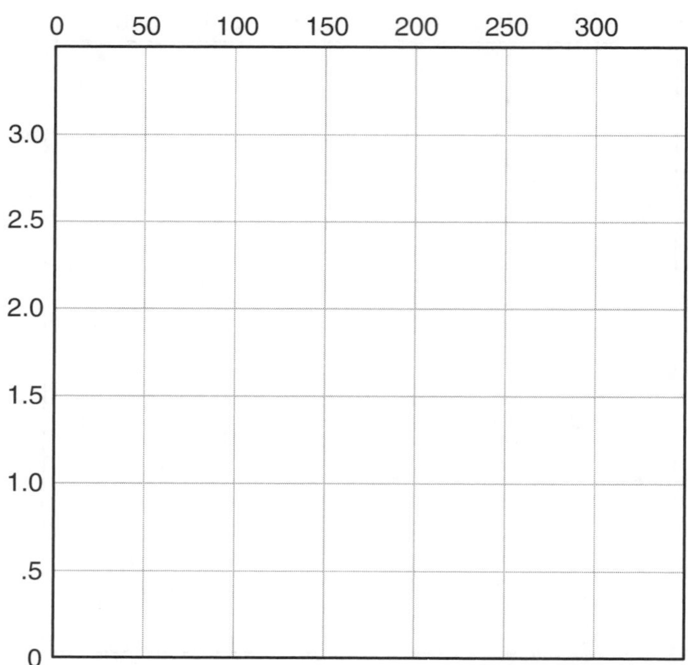

5. What percentage of full-flow pressure is the cracking pressure for the each of the four valve test procedures?

   cracking pressure % = cracking pressure ÷ valve pressure setting

   A. Direct-acting relief valve

      200 psi setting: ______________________________

      300 psi setting: ______________________________

   B. Pilot-operated relief valve

      200 psi setting: ______________________________

      300 psi setting: ______________________________

6. Explain why the cracking pressure percentage increases as the maximum pressure setting of the valves increases.

____

____

____

____

7. Explain the reasons for the variations in the percentages between the four calculated cracking percentages.

____

____

____

____

8. Explain the term *pressure override* as it relates to pressure control valves.

9. Why is an understanding of the concepts of cracking pressure and pressure override important when selecting and installing a relief valve in a hydraulic system?

10. How do the concepts of cracking pressure, pressure override, and full-flow pressure relate to the operation of other control valves, such as sequence and counterbalance valves?

Activity 10-3

Identify and Compare the Operating Characteristics of Open-, Closed-, and Tandem-Center Four-Way Directional Control Valves

Name ______________________________ Date ______________

This activity illustrates the operating concepts involved with the three most common center configurations used with three-position, four-way directional control valves. The activity illustrates the effects these center position designs have on the performance of the pump and the system actuator. Although many additional center designs are available, an understanding of these three basic designs illustrates how directional control valves can be used to obtain maximum system operating efficiency and motion control.

Activity Specifications

Set up a test circuit to illustrate the flexibility provided by the center position of a four-way directional control valve. Compare system fluid pressure and flow rates for each of the center designs with the directional control valve in the center position and during actuator movement. Use the test circuit and procedure shown below to test the performance using each center configuration. Compare the test results to establish the desirability of each configuration for hydraulic circuit applications.

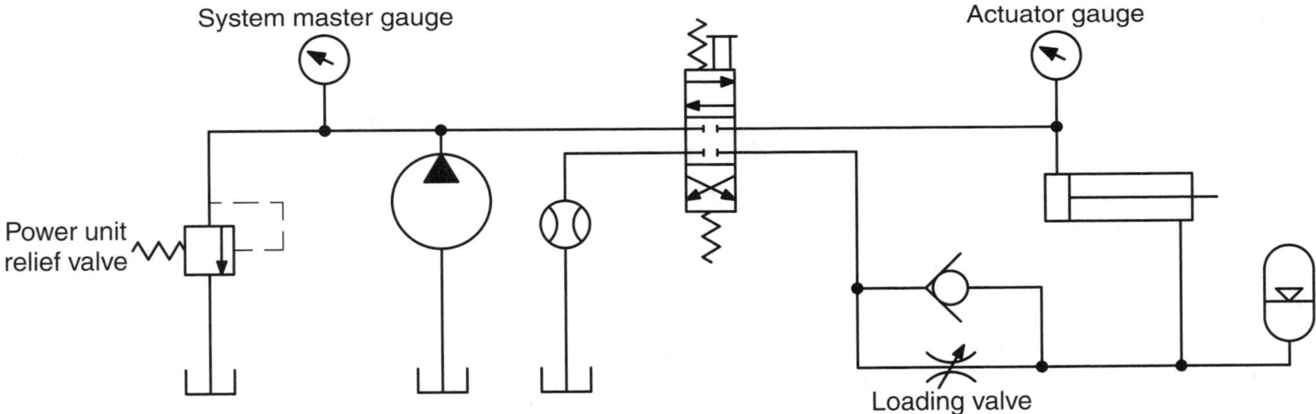

Required Components

Access to the following components is needed to complete this activity.

- Hydraulic power unit (pump, reservoir, and prime mover).
- Power unit relief valve.
- Two pressure gauges (one used as the power unit master pressure gauge and the second as the actuator pressure gauge).
- Flow control valve with built-in check valve (used as the loading device for the circuit actuator).
- Accumulator (used in conjunction with the loading device to maintain a load on the circuit actuator when the directional control valve is shifted to the center position).
- Flowmeter (used to identify the flow returned to the reservoir through the directional control valve).

- Three three-position, four-way directional control valves (used as test valves; one valve with a closed center, a second valve with an open center, and a third valve with a tandem center).
- Manifolds (adequate number to make gauge connections).
- Hoses (adequate number to allow assembly of the system).

Procedures

Study the following procedures to become familiar with the steps needed to complete the activity. Then, complete the procedures and record the observations indicated in the Data and Observations Record section.

1. Select the components needed to assemble the system and test circuit.
2. Have your instructor check your selection of components.
3. Under the *supervision* of your instructor, charge the accumulator with inert gas to 100 psi.

Closed-center directional control valve

4. Assemble the system using the closed-center directional control valve. Have your instructor check the setup of the circuit before proceeding.
5. With the directional control valve in the centered position, adjust the power unit relief valve until the master pressure gauge reads 400 psi.
6. Completely open the loading valve of the test circuit. This setting represents a no-load condition on the system.
7. Shift the directional control valve to extend the cylinder. Note the master gauge pressure, actuator gauge pressure, and fluid flow through the flowmeter as the cylinder is *extending*. Record these readings in the Data and Record Observations section.
8. Shift the directional control valve to retract the cylinder.
9. Completely extend the cylinder. Momentarily hold the directional control valve in this shifted position before shifting the valve into the center position.
10. Note and record the master gauge pressure, actuator gauge pressure, and flow through the flowmeter when the valve is in the *centered* position.
11. Shift the directional control valve to retract the cylinder.
12. By experimentation, close the loading valve until the flow rate through the flowmeter is 1/2 of the flow rate observed through the meter in step 7.
13. Repeat steps 7 through 11 to observe the operation of the directional control valve under partial-load operating conditions.
14. Completely close the loading valve.
15. Repeat steps 7 through 11 to observe the operating characteristics of the closed-center directional valve when the system actuator is stalled by a system load.

Open-center directional control valve

16. Replace the closed-center directional control valve with the open-center valve.
17. Completely open the loading valve of the test circuit. This setting represents a no-load condition on the system.
18. Shift the directional control valve to extend the cylinder. Note the master gauge pressure, actuator gauge pressure, and fluid flow through the flowmeter as the cylinder is *extending*. Record these readings in the Data and Record Observations section.
19. Shift the directional control valve to retract the cylinder.

20. Completely extend the cylinder. Momentarily hold the directional control valve in this shifted position before shifting the valve into the center position.
21. Note and record the master gauge pressure, actuator gauge pressure, and flow through the flowmeter when the valve is in the *centered* position.
22. Shift the directional control valve to retract the cylinder.
23. By experimentation, close the loading valve until the flow rate through the flowmeter is 1/2 of the flow rate observed through the meter in step 18.
24. Repeat steps 18 through 22 to observe the operation of the open-center directional control valve under partial-load operating conditions.
25. Completely close the loading valve.
26. Repeat steps 18 through 22 in this section to observe the operating characteristics of the closed-center valve when the system actuator is stalled by a system load.

Tandem-center directional control valve

27. Replace the open-center directional control valve with the tandem-center valve.
28. Shift the directional control valve to extend the cylinder. Note the master gauge pressure, actuator gauge pressure, and fluid flow through the flowmeter as the cylinder is *extending*. Record these readings in the Data and Record Observations section.
29. Shift the directional control valve to retract the cylinder.
30. Completely extend the cylinder. Momentarily hold the directional control valve in this shifted position before shifting the valve into the center position.
31. Note and record the master gauge pressure, actuator gauge pressure, and flow through the flowmeter when the valve is in the *centered* position.
32. Shift the directional control valve to retract the cylinder.
33. By experimentation, close the loading valve until the flow rate through the flowmeter is 1/2 of the flow rate observed through the meter in step 29.
34. Repeat steps 29 through 33 to observe the operation of the directional control valve under partial-load operating conditions.
35. Completely close the loading valve.
36. Repeat steps 29 through 33 to observe the operating characteristics of the tandem-center valve when the system actuator is stalled by a system load.
37. Stop the power unit and have your instructor check the results of your tests.
38. Disassemble the circuit; clean and replace components in their assigned location.

Data and Observation Records

Before conducting the laboratory test procedures, use manufacturer catalogs to collect published information about the directional control valves used in this activity. Record the basic manufacturer information requested. Then, carefully record the pressure and flow information from the test procedures in the charts provided.

Component information

	Closed-Center Valve	Open-Center Valve	Tandem-Center Valve
Brand			
Model number			
Pressure rating			
Flow capacity			

Test data

Valve Center Design	System Load Condition	Directional Valve Position	Master Gauge psi	Actuator Gauge psi	Flowmeter gpm
Closed	No load	Extending			
		Centered			
	Partial load	Extending			
		Centered			
	Stalled	Extending			
		Centered			
Open	No load	Extending			
		Centered			
	Partial load	Extending			
		Centered			
	Stalled	Extending			
		Centered			
Tandem	No load	Extending			
		Centered			
	Partial load	Extending			
		Centered			
	Stalled	Extending			
		Centered			

Activity Analysis

Answer the following questions in relation to the collected data and observations of the test procedures.

1. Why does the actuator gauge register pressure when the closed- and tandem-center directional control valves are shifted into the center position? What is the source of this pressure and what does it represent in this activity circuit?

2. Compare the master gauge pressure readings when the open-, closed-, and tandem-center valves are centered. What are the reasons for the pressure differences that exist among the valves?

3. Examine the master gauge pressure readings and the flow rates through the flowmeter when the open- and tandem-center valves are in the centered position. Explain the reasons for any variations.

4. Compare the actuator gauge pressure readings of the three valve center designs when the cylinder is being extended under a no-load condition. Why is there little or no difference between the readings?

____

____

____

____

5. Compare the flowmeter flow readings when each of the three directional control valves is centered. Explain any variations in the flow rates.

____

____

____

____

6. Which of the three valve center position designs allows the pump prime mover to consume the least amount of energy while the directional control valve is center? Why? Use the data collected during the activity to support your answer.

____

____

____

____

7. Describe a simple application for each of the three directional control valve centers that would allow the characteristics of the design to be used to the best advantage.

____

____

____

____

____

____

Activity 10-4

Test and Compare the Performance of Noncompensated and Compensated Flow Control Valves

Name ______________________ **Date** ____________

This activity illustrates the performance differences of the two commonly used types of flow control valves. The activity illustrates a simple test circuit and procedure that can be used to show the ability of noncompensated and compensated flow control valves to respond to varying actuator loads. Follow-up questions requiring the analysis of the collected data not only require a comparison of the two flow control valves types, but also an awareness of relief valve cracking pressure. This pressure must be considered to assure an efficiently operated system.

Activity Specifications

Set up a test circuit to measure the performance of two types of hydraulic flow control valves. Use the test circuit and procedure below to compare the performance of a simple noncompensated and a compensated flow control valve. Plot and compare the performances of these valves as the system load is varied in the test circuit. Relate test results to the operation of flow control valves in an operating hydraulic circuit.

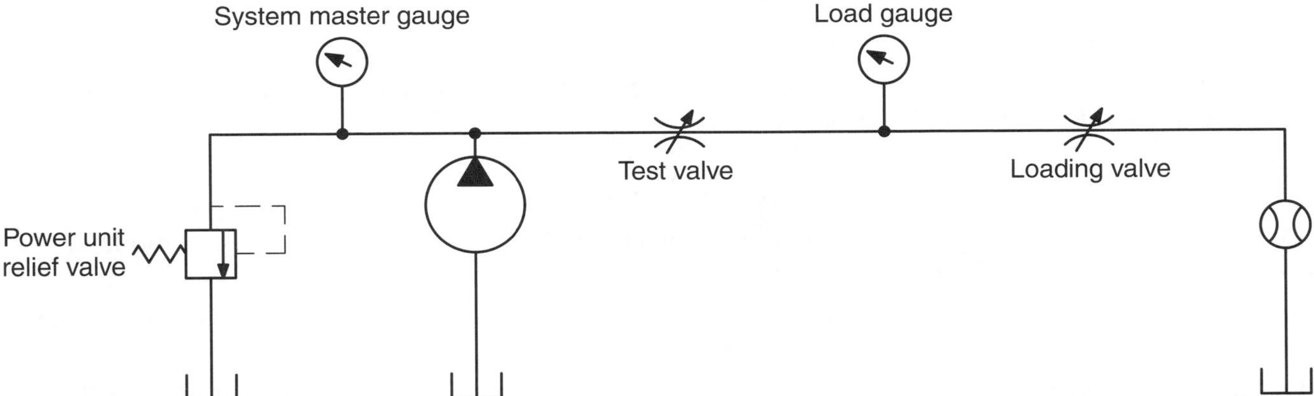

Required Components

Access to the following components is needed to complete this activity.

- Hydraulic power unit (pump, reservoir, and prime mover).
- Power unit relief valve.
- Two pressure gauges (one used as the system master pressure gauge and the second as the load pressure gauge).
- Two noncompensated flow control valves (one used as one of the test valves in the activity and the second as the loading device for testing the valves).
- One compensated flow control valve (used as one of the test valves in the activity).
- Flowmeter (used to identify the flow discharged to the reservoir through the test flow control valves).
- Manifolds (adequate number to make gauge connections).
- Hoses (adequate number to allow assembly of the system).

Procedures

Study the following procedures to become familiar with the steps needed to complete the activity. Then, complete the procedures and record the observations indicated in the Data and Observations Record section.

1. Select the components needed to assemble the system and test circuit.
2. Collect and record catalog data on the two flow control valves being tested in this activity.
3. Assemble the system using the noncompensated flow control as the test valve.
4. Have your instructor check your selection of components and the setup of the circuit before proceeding.

Set the initial pressure and flow settings of the circuit valves

5. Adjust the system relief valve to the lowest-possible pressure setting.
6. Completely open the test and loading valves.
7. Start the power unit.
8. Completely close the loading valve. Adjust the system relief valve until the system pressure gauge reads 500 psi.
9. Completely open the loading valve of the test circuit.
10. Record the initial pressures and flow rate of the test circuit in the Data and Observation Records section.

Establish initial flow rate of test flow control valve

11. Close the test valve until the flowmeter reads 1 gpm.
12. Record the pressure and flow readings in the chart for the noncompensated flow control valve in the Data and Observation Records section.

Test the valves under varying load conditions

13. Slowly close the loading valve until the load gauge reads 100 psi. Record the system and load pressures and the flowmeter readings.
14. Continue to close the loading valve, recording pressure and flow readings at 50 psi intervals, until the load gauge reaches the maximum pressure.
15. Check the reading taken by slowly opening the loading valve. Check the readings again as the gauge pressure is reduced by 50 psi intervals.
16. Stop power unit.
17. Replace the noncompensated flow control valve with the compensated valve.
18. Completely open the test and loading valves.
19. Start the power unit.
20. Completely close the loading valve. Adjust the system relief valve until the system pressure gauge reads 500 psi.
21. Completely open the loading valve of the test circuit.
22. Record the initial pressures and flow rate of the test circuit in the Data and Observation Records section.
23. Close the test valve until the flowmeter reads 1 gpm.
24. Record the pressure and flow readings in the chart for the noncompensated flow valve in the Data and Observation Records section.
25. Slowly close the loading valve until the load gauge reads 100 psi. Record the system and load pressures and the flowmeter readings.

26. Continue to close the loading valve, recording pressure and flow readings at 50 psi intervals, until the load gauge reaches the maximum pressure.
27. Check the reading taken by slowly opening the loading valve. Check the readings again as the gauge pressure is reduced by 50 psi intervals.
28. Stop power unit.
29. Record all data in the chart for the compensated flow control valve in the Data and Observation Records section.
30. Stop the power unit and have your instructor check the results of your tests.
31. Disassemble the circuit; clean and replace components in their assigned location.

Data and Observation Records

Before conducting the laboratory test procedures, use manufacturer catalogs to collect information about the flow control valves selected for this activity. Record the basic information requested for use in analysis of the valve test results. Then, carefully record the pressure and flow information from the test procedures in the charts provided.

Component information

	Noncompensated Valve	Compensated Valve
Brand		
Model number		
Pressure rating		
Flow capacity		

Noncompensated flow control valve test

A. Initial pressure gauge and flowmeter readings

System pressure gauge: ______________________

Load pressure gauge: ______________________

Flowmeter: ______________________

B. Test data

Load gauge psi	System gauge psi	Flowmeter gpm
		1
100		
150		
200		
250		
300		
350		
400		
450		
500		

Compensated flow control valve test

A. Initial pressure gauge and flowmeter readings

System pressure gauge: ______________________________

Load pressure gauge: ______________________________

Flowmeter: ______________________________

B. Test data

Load gauge psi	System gauge psi	Flowmeter gpm
		1
100		
150		
200		
250		
300		
350		
400		
450		
500		

Plots of Valve Performance

Use the information collected in the Data and Observation Records section to plot the performance of the valves.

Noncompensated flow control valve

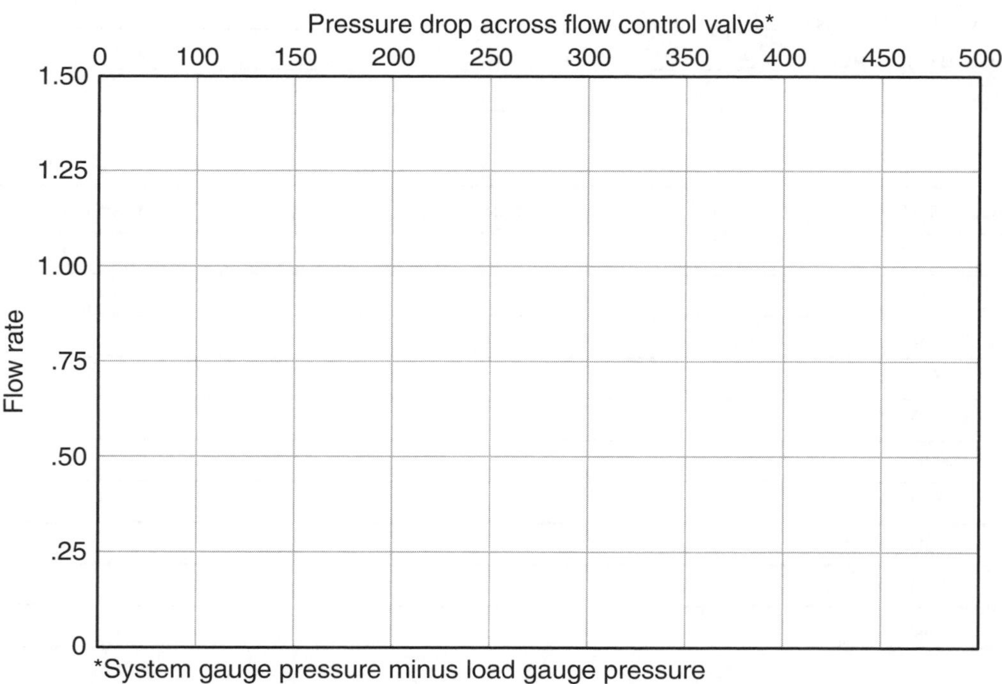

Compensated flow control valve

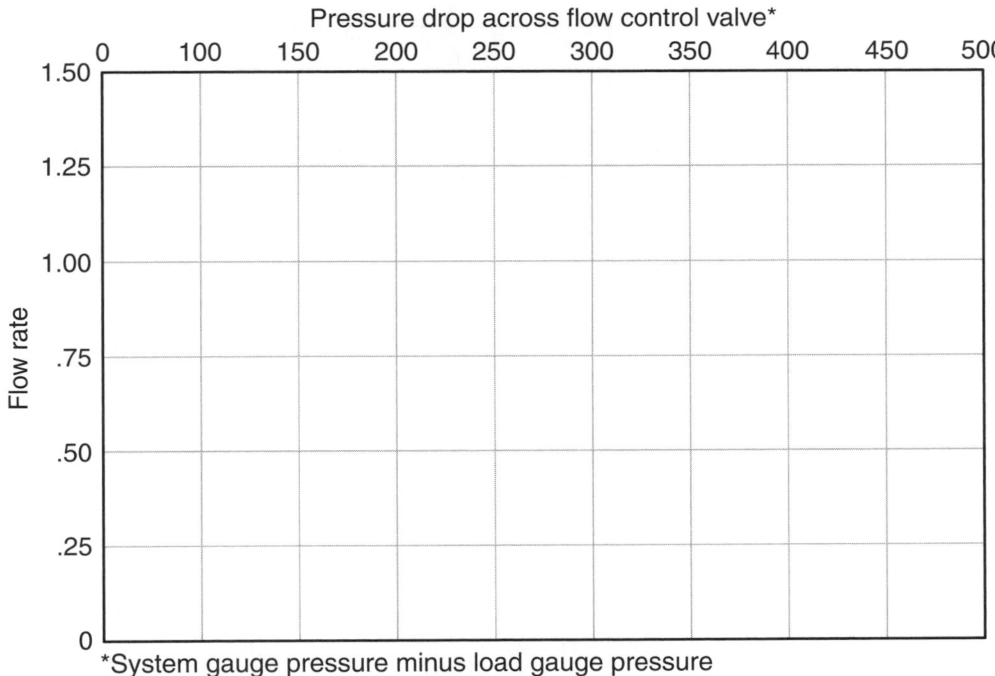

Activity Analysis

Answer the following questions in relation to the collected data and observations of the test procedures.

1. Based on the plots, what is the primary performance difference between the two basic flow control valve designs?

2. What does the loading valve represent in this test circuit?

3. Why does the flow through the noncompensated flow control valve vary as the system load changes?

4. Explain how the mechanism of the compensated flow control valve maintains a uniform fluid flow rate through the valve under varying load conditions.

5. Under what type of circuit load conditions would it be adequate to use a noncompensated flow control valve to maintain uniform actuator speed?

Chapter 11

Accumulators

Pressure, Flow, and Shock Control Assistance

The activities in this lab exercise are designed to show several accumulator designs and application factors, as discussed in the chapter. Emphasis is placed on demonstrating those concepts key to effective and safe use of accumulators in hydraulic systems. The activities demonstrate basic principles relating to accumulator sizing, the volume of fluid available from an accumulator for circuit operation under various precharge pressures, and a simple holding circuit using an accumulator.

Caution: Check with your instructor to be certain you are working with the correct equipment. Also, follow all safety procedures for your laboratory as you complete the activities.

Key Terms

The following terms are used in this chapter. As you read the text, record the meaning and importance of each. Additionally, you may use other sources, such as manufacturer literature, an encyclopedia, or the Internet, to obtain more information.

accumulators ______________________

____

____

adiabatic ______________________

____

____

gas-charged accumulators ______________________

____

____

isothermal ______________________

____

____

precharging ______________________

____

____

pressure rating ______________________

____

____

pressure surges ______________________

____

____

rated capacity ______________________

____

____

shock pressure ______________________

____

____

spring-loaded accumulator ______________________

____

____

usable volume ______________________

____

____

weight-loaded accumulator ______________________

____

____

Chapter 11 Quiz

Name ______________________________ Date ____________ Score ________

Write the best answer to each of the following questions in the blanks provided.

1. List the three methods used to store energy in an accumulator.

____

____

____

2. Which is the only accumulator design that can deliver their entire fluid volume at a near-constant pressure?

____

3. Why does a spring-loaded accumulator not provide a constant rate of fluid intake or discharge?

____

____

____

____

4. ______ is the gas normally used to precharge gas-charged accumulators.

____

5. Name the two chambers involved in the structure of all gas-charged accumulators.

____

____

____

6. During hydraulic system operation, which chamber of a gas-charged accumulator is subjected to actual system pressure?

____

7. A gas-charged accumulator begins to store fluid and energy when system pressure exceeds the ______ pressure.

____

8. What three factors affect a gas, as described by the ideal gas laws?

____

____

____

9. A(n) ______ condition exists when an accumulator is filled over an extended time period since there is no temperature change.

____

10. List the four ways in which accumulators can increase the efficiency of a hydraulic system.

____

____

____

____

11. What are two factors that can justify the use of an accumulator to supplement pump output flow?

____

____

12. Accumulators are available with rated maximum operating pressures ranging from 2000 psi to over ______ psi.

____

13. The ideal mounting position for any accumulator is ______.

____

14. What is often applied around fittings and seams of gas-charged accumulators to assist in detecting external gas leaks?

____

15. Before removing an accumulator from a hydraulic system, the pressure in the unit should be reduced to ______ psi.

____

Activity 11-1

Select an Accumulator Using Manufacturer Catalog Data

Name ______________________________ **Date** ______________

This activity is designed to show how a fluid power component manufacturer presents technical information about the gas-charged accumulators it designs, manufactures, and markets. Technical data includes information relating to the physical size of components and the volume of fluid available for applications involving various precharge and system operating pressures. Many manufacturers also supply considerable information to use when selecting the correct-capacity accumulator for an application.

Activity Specifications

Analyze the data sheets and apply selection guidelines for a gas-charged hydraulic accumulator. Section One of the activity requires a review of data sheets to identify the basic information available about an accumulator. Become familiar with the data sheets before attempting to answer the activity questions. Section Two of the activity involves the sizing of an accumulator using selection guidelines and hydraulic circuit parameters provided by your instructor.

Section One—Data Sheet Review

A number of manufacturers produce several types and sizes of accumulators. Typically, these companies supply considerable information about the construction, capacity, and maintenance of their units. This information is usually available in either a printed catalog or on the company's website. This portion of the activity should be completed using one of these sources for the accumulator brand and model assigned by your instructor.

Assigned accumulator

Accumulator type: ______________________________

Manufacturer: ______________________________

Model number: ______________________________

Data identification and analysis

1. Identify the maximum recommended operating pressure for the accumulator. In your judgment, what design and construction factors were considered in establishing this rating?

____

____

____

____

2. What is the storage capacity of the accumulator? What is the maximum recommended flow rating of the unit? What is the difference between these two ratings?

____

____

____

____

3. Identify the height and diameter of the accumulator. What other dimensions must be considered when selecting an accumulator for a specific application?

____

____

____

____

4. What is the style and size of the hydraulic port on the accumulator? How does port size relate to the performance of the accumulator when applied to a specific application?

____

____

____

____

5. Which components in an accumulator can be adversely affected by the type of fluid used in the system? List the various options available for these parts and the fluids with which they can be safely used.

____

____

____

____

6. If the accumulator is gas charged, describe the type of gas valve used for charging it. Briefly explain the recommended method for changing the precharge pressure.

____

____

____

____

Section Two—Sizing an Accumulator

Manufacturers typically provide formulas for determining the size of an accumulator for use in a hydraulic system application. In addition, these companies often provide direct technical support to assure the selection of the correct type and capacity of accumulator for an application. The following websites provide formulas or calculators that may be used in sizing accumulators for various applications. Review these sites or similar sites recommended by your instructor before continuing with this portion on the activity.

www.accumulators.com

www.wilkesandmclean.com

The following section of the activity involves determining the size of an accumulator for two common accumulator applications. Obtain the basic parameters of the applications from your instructor. Use a suggested formula or calculator from a website or other source to complete the following problems.

Energy storage application

Determine the standard accumulator size required to meet the following system storage conditions.

Oil volume required from the accumulator: ____________________________________

Maximum system pressure: ____________________________________

Minimum system pressure: ____________________________________

Accumulator charge time (seconds): ____________________________________

Accumulator discharge time (seconds): ______________________

Recommended accumulator capacity: ______________________

Leakage compensator application

Determine the standard accumulator size required to meet the following system leakdown conditions.

Maximum system pressure: ______________________

Minimum system pressure: ______________________

System leakage rate (cubic inches per minute): ______________________

System hold time (minutes): ______________________

Recommended accumulator capacity: ______________________

Activity Analysis

Answer the following questions in relation to the information found on the recommended websites and the calculations completed in the activity.

1. What types of services are offered by the accumulator manufacturer through their website? Name at least two reasons for these offers of assistance.

2. How many different accumulator applications are included in the formulas and calculators on the various websites? How similar are the approaches used by the companies? Why?

3. Which scale should be used when specifying the pressure in the formulas? Why?

4. How important is precharge pressure in these various calculations?

5. Why must the size of fittings be carefully considered when connecting an accumulator to the system lines?

6. Are adiabatic and isothermal heat transfer principles involved in any of these calculations? Explain your answer.

Activity 11-2

Test the Storage Capacity of an Accumulator under Varying Precharge and System Operating Conditions

Name ______________________________ **Date** ______________

The volume of hydraulic oil that can be stored in an accumulator depends on precharge gas pressure and system pressure. Manufacturers typically provide charts listing the storage capacity of their accumulator models under various precharge and system load conditions. This activity is designed to verify the storage capacity of a gas-charged accumulator under varying system loads to provide a better understanding of the operating principles of accumulators.

Activity Specifications

Set up a test circuit to measure the performance of a gas-charged hydraulic accumulator. Measure the volume of hydraulic fluid stored in the accumulator with a fixed precharge pressure and varying system pressures. Plot the performance of the accumulator and compare it with the manufacturer's published data. Use the test circuit and procedure shown below to test the accumulator.

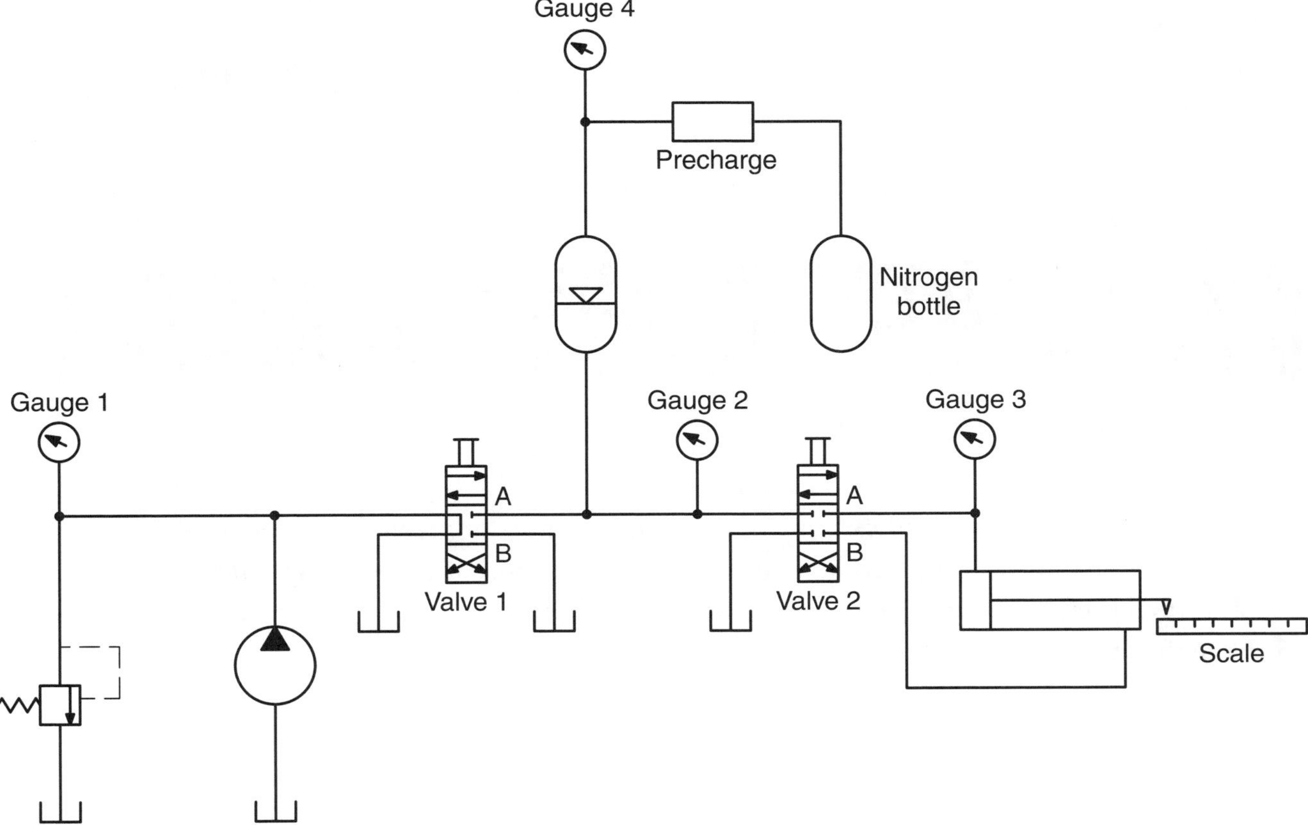

Required Components

Access to the following components is needed to complete this activity.

- Hydraulic power unit (pump, reservoir, and prime mover).
- Power unit relief valve.
- Four pressure gauges (one used as the power unit master pressure gauge and the others to measure pressure at test points in the circuit).
- Two three-position, four-way directional control valves (one with a closed center and the other an open center).
- Accumulator (piston type is preferred).
- Accumulator precharge fittings and gauging components.
- Nitrogen-charged gas bottle (appropriate for precharge equipment).
- Double-acting cylinder (used as a volume-measuring device).
- Scale (used to measure cylinder travel).
- Manifolds (adequate number to make gauge connections).
- Hoses (adequate number to allow assembly of the system).

Procedures

Study the following procedures to become familiar with the steps needed to complete the activity. Then, complete the procedures and record the observations indicated in the Data and Observation Records section.

1. Select the components needed to assemble the system and test circuit. Identify manufacturer catalog data for the accumulator.
2. Assemble the system.
3. Have your instructor check your selection of components and the setup of the circuit before proceeding.

Caution: Do *not* continue with this activity until your instructor reviews the precharge procedure and related safety factors and has given *permission* to continue with the activity.

Precharge the accumulator

1. Shift directional control valve 1 into position B.
2. Hold the valve in the shifted position during the accumulator precharge procedure.
3. Follow the manufacturer's recommended procedure to precharge the accumulator to 200 psi.
4. Shift directional control valves 1 and 2 into the center position.
5. Note the initial, retracted position of the cylinder rod.

Measure accumulator storage capacity under various conditions

1. Start the hydraulic power unit.
2. Shift directional control valve 1 to position A to pressurize the accumulator section of the test circuit.
3. Adjust the system relief valve to produce a system pressure of 150 psi, as indicated on gauge 1. Note and record the pressures shown on gauges 2, 3, and 4 in chart 1 in the Data and Observation Records section.
4. Shift directional control valve 1 to the center position. Note and record the pressure on all gauges.
5. Shift directional control valve 2 to position A. Measure the distance the cylinder extends and record it in chart 1.
6. Note the pressure registered on all circuit gauges while holding directional control valve 2 in position A. Record these pressures in chart 1.
7. Simultaneously shift directional control valve 1 into position A and valve 2 into position B to retract the cylinder.
8. Repeat steps 2 through 7 for each of the additional system pressures shown in chart 1. These pressures should be read on gauge 1.
9. Readjust the precharge pressure of the accumulator to 300 psi.
10. Repeat steps 2 through 7 for the system pressures shown in chart 2.

Activity completion

1. Discuss your data and observations with your instructor before disassembling the test circuit.
2. Clean and return all component parts to their assigned storage locations.
3. Complete the activity questions.

Data and Observation Records

Before conducting the laboratory test procedures, use manufacturer catalogs to collect information about accumulator used in the activity. Record the basic information below. Then, carefully record the pressure and cylinder movement information requested at various points during the test procedures.

Accumulator information

Basic design: ____________________

Manufacturer: ____________________

Model number: ____________________

Fluid capacity: ____________________

Pressure rating: ____________________

Chart 1—precharge pressure 200 psi

System Pressure Gauge 1	Directional Control		Pressure			Cylinder Rod Travel
	Valve #	Spool Position	Gauge 2	Gauge 3	Gauge 4	
150	1	Centered				
	2	Centered				
	1	A				
	2	Centered				
	1	Centered				
	2	A				
200	1	Centered				
	2	Centered				
	1	A				
	2	Centered				
	1	Centered				
	2	A				
250	1	Centered				
	2	Centered				
	1	A				
	2	Centered				
	1	Centered				
	2	A				
300	1	Centered				
	2	Centered				
	1	A				
	2	Centered				
	1	Centered				
	2	A				
350	1	Centered				
	2	Centered				
	1	A				
	2	Centered				
	1	Centered				
	2	A				

Chart 2—Precharge pressure 300 psi

System Pressure Gauge 1	Directional Control		Pressure			Cylinder Rod Travel
	Valve #	Spool Position	Gauge 2	Gauge 3	Gauge 4	
250	1	Centered				
	2	Centered				
	1	A				
	2	Centered				
	1	Centered				
	2	A				
300	1	Centered				
	2	Centered				
	1	A				
	2	Centered				
	1	Centered				
	2	A				
350	1	Centered				
	2	Centered				
	1	A				
	2	Centered				
	1	Centered				
	2	A				
400	1	Centered				
	2	Centered				
	1	A				
	2	Centered				
	1	Centered				
	2	A				
450	1	Centered				
	2	Centered				
	1	A				
	2	Centered				
	1	Centered				
	2	A				

Storage Capacity Calculation and Plots

The test circuit in this activity uses the displacement of a cylinder to measure the amount of fluid stored under various circuit conditions. This section requires calculation of the storage capacity under each of the activity test conditions. Simple plots are then produced to illustrate the effect of precharge and system pressures on the actual volume of oil stored in an operating accumulator.

1. Calculate the volume of fluid discharged by the accumulator for each of the system pressure settings tested in charts 1 and 2.

   Cylinder bore: ____________________

   Rod diameter: ____________________

200 psi precharge

System Pressure (psi)	Cylinder Travel (inches)	Accumulator Discharge (cubic inches)
150		
200		
250		
300		
350		

300 psi precharge

System Pressure (psi)	Cylinder Travel (inches)	Accumulator Discharge (cubic inches)
250		
300		
350		
400		
450		

2. Plot the fluid discharge of the accumulator for the 200 psi and 300 psi precharges. Establish the low-to-high storage scale portion of the charts based on data calculated in question 1.

200 psi precharge

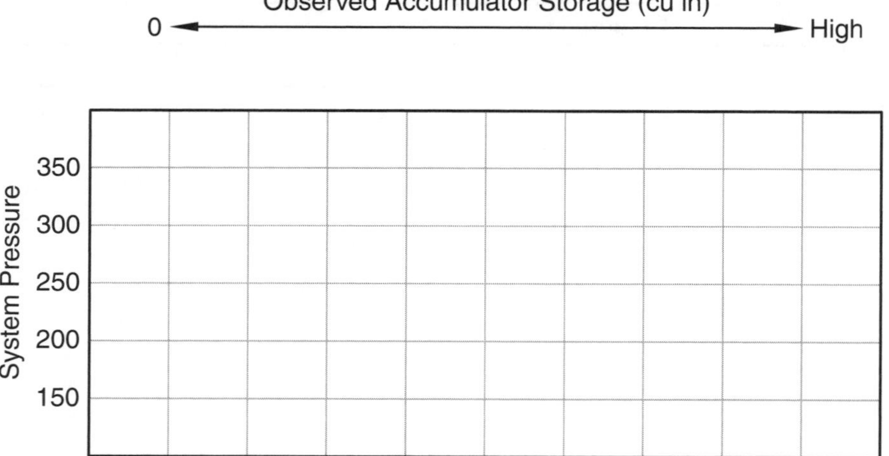

300 psi precharge

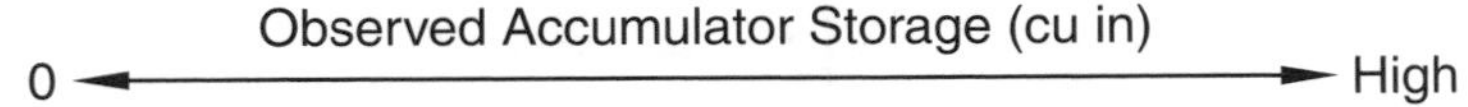

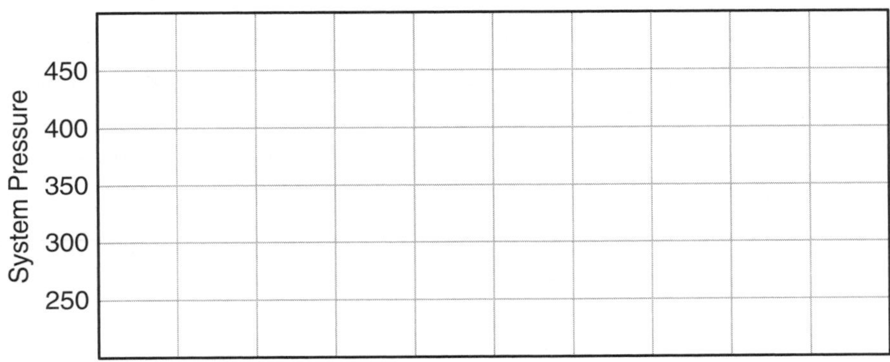

Activity Analysis

Complete the following in relation to the data collected and observations of the test procedures.

1. List the safety precautions that need to be followed during the precharging of a gas-charged accumulator. Consult manufacturer literature and the chapter material.

2. Why is it *critical* to use nitrogen gas to precharge an accumulator?

3. Why is it necessary to be certain the accumulator is fully discharged before the precharge procedure is implemented?

4. At what point in the test procedure does the accumulator begin to store energy? Indicate the accumulator gas pressures at which this occurs in the two test procedures performed in this activity.

5. Identify any similarities or differences between the plots for the two precharge pressures developed in this activity. Discuss the reasons for any similarities or differences.

6. Do the capacity figures obtained in this activity result from isothermal or adiabatic compression of the precharge gas? Explain the rational for your answer.

Activity 11-3

Test the Operation of a Circuit Using an Accumulator to Maintain Pressure

Name ______________________________ Date ______________

A gas-charged accumulator allows a hydraulic system to maintain the advantages of a noncompressible liquid while using a gas to enhance system operation. The efficiency of a hydraulic system can be increased by using an accumulator to store energy, help control system pressure and flow variations, maintain system pressure, and supplement pump output. This activity uses simple circuits to illustrate the basic principles of maintaining pressure.

Activity Specifications

Construct and test two basic circuits that provide three operating phases for a hydraulic-powered press cylinder: extension, holding, and retraction. Measure the pressures involved in the cylinder and pump sections of the circuits during each of the three operating phases. Use the test circuits and the procedures shown below to set up and operate the circuits. Record the test data, compare the operating pressures between the circuits, and answer the questions.

Circuit 1

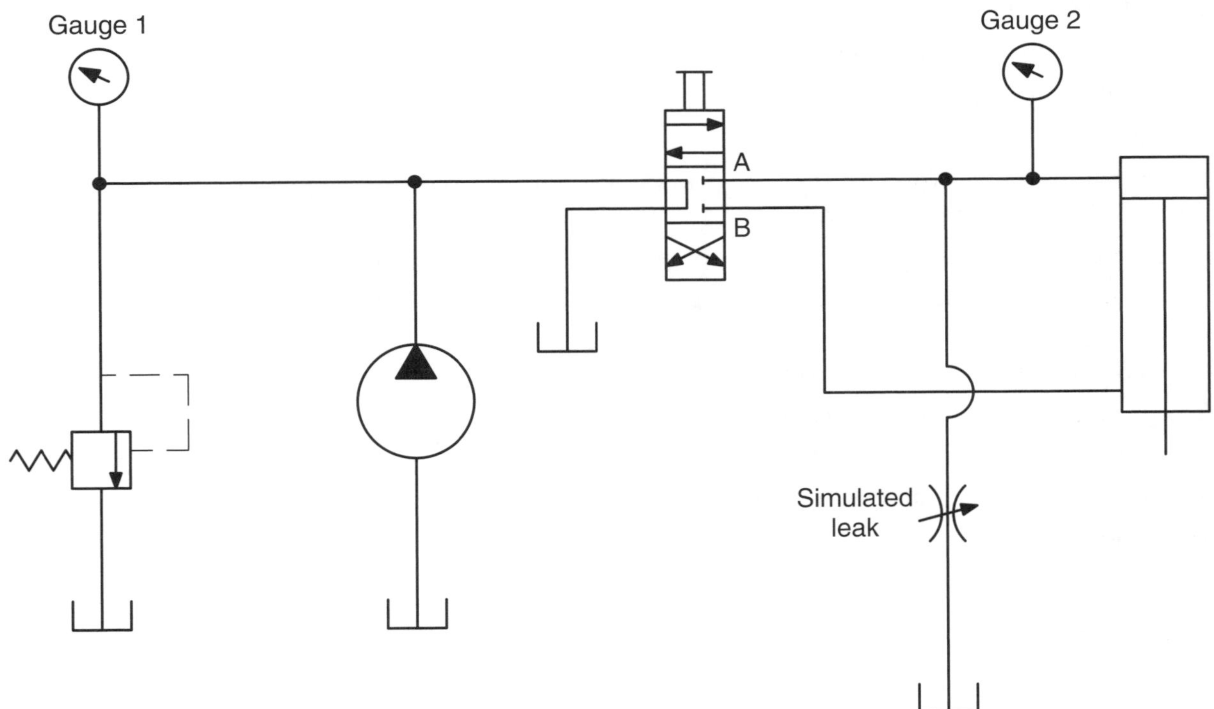

Circuit 2

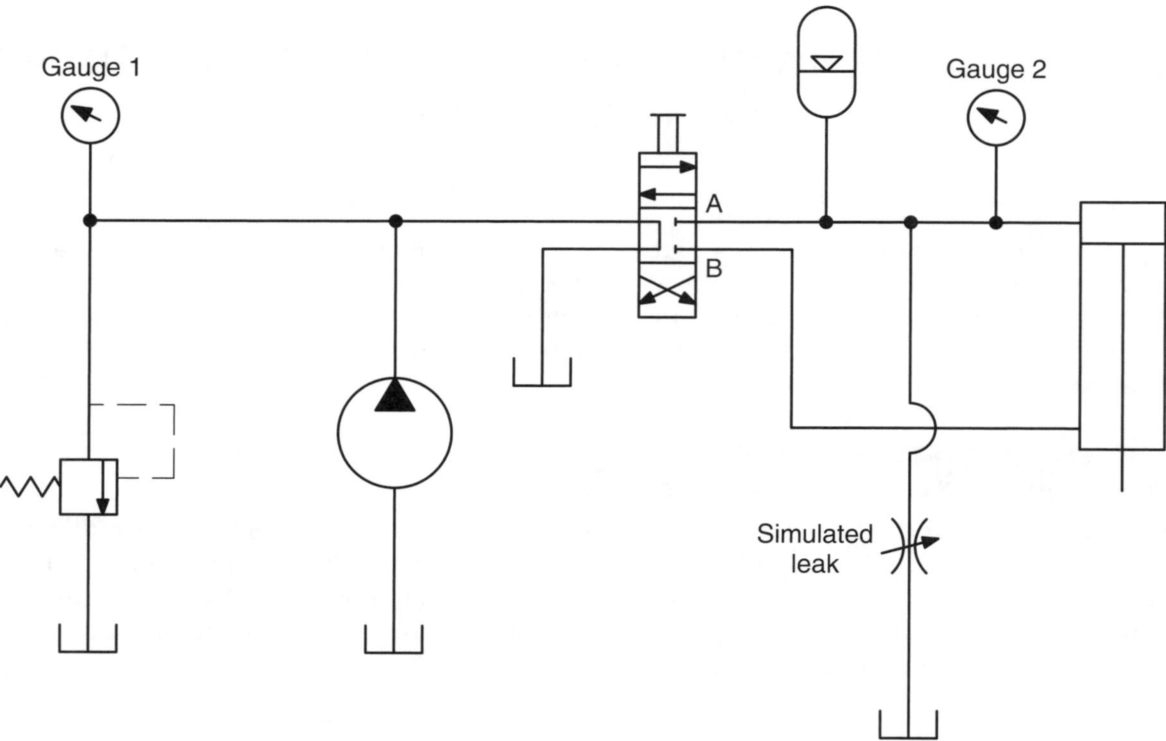

Required Components

Access to the following components is needed to complete this activity.

- Hydraulic power unit (pump, reservoir, and prime mover).
- Power unit relief valve.
- Two pressure gauges (one used as the power unit master pressure gauge and the other to measure cylinder working pressure).
- Three-position, four-way directional control valve (tandem center).
- Double-acting cylinder (large-diameter unit with a relatively short stroke to simulate a press actuator).
- Flow control valve (noncompensated needle valve).
- Accumulator (piston type is preferred).
- Accumulator precharge fittings and gauging components.
- Nitrogen-charged gas bottle (appropriate for precharge equipment).
- Manifolds (adequate number to make gauge connections).
- Hoses (adequate number to allow assembly of the system).

Procedures

Study the following procedures to become familiar with the steps needed to complete the activity. Then, complete the procedures and record the observations indicated in the Data and Observation Records section.

1. Select the components needed to assemble the system and test circuit.
2. Assemble circuit 1. Have your instructor check your selection of components and the setup of the circuit before proceeding.
3. Completely open the system relief valve and start the power unit.
4. Adjust the needle valve to allow a very small volume of hydraulic fluid to bleed to the reservoir. This valve is used to simulate the internal leakage that can occur in an actual operating system.
5. Adjust the relief valve to set the maximum system operating pressure at 300 psi.
6. Shift the directional control valve to fully extend the cylinder.
7. Hold the directional control valve in this shifted position to maintain maximum pressure throughout the system.
8. Note the pressures on gauges 1 and 2. Record these pressures in chart 1 in the Data and Observation Records section.
9. Shift the directional control valve into the center position. Note and record the pressures on gauges 1 and 2 immediately after the valve has been shifted.
10. After 30, 60, and 90 seconds, again note and record the pressures on the gauges.
11. Shift the directional control valve to fully retract the cylinder. Hold the valve in the shifted position.
12. Note and record the pressures on gauges 1 and 2.
13. Repeat steps 5 through 12 for system operating pressures of 400 psi and 500 psi.
14. Stop the power unit and insert an accumulator as shown in circuit 2.

Caution: Do *not* continue with this activity until your instructor reviews the precharge procedure and related safety factors and has given permission to continue with the activity.

15. Follow the recommended procedure to precharge the accumulator to 200 psi.
16. Repeat steps 5 through 13, recording your observations in chart 2.
17. Discuss your data and observations with your instructor before disassembling the test circuit.
18. Clean and return all components to their assigned storage locations.

Data and Observation Records

Carefully note and record the information requested at various points during test procedures.

Record chart 1—without accumulator

Relief Valve Setting (psi)	Directional Control Valve	Observation Point	Pressure	
			Gauge 1	Gauge 2
300	Position A	Extended		
	Centered	Immediate		
		30 seconds		
		60 seconds		
		90 seconds		
	Position B	Retracted		
400	Position A	Extended		
	Centered	Immediate		
		30 seconds		
		60 seconds		
		90 seconds		
	Position B	Retracted		
500	Position A	Extended		
	Centered	Immediate		
		30 seconds		
		60 seconds		
		90 seconds		
	Position B	Retracted		

Record chart 2—accumulator precharge 200 psi

Relief Valve Setting (psi)	Directional Control Valve	Observation Point	Pressure	
			Gauge 1	Gauge 2
300	Position A	Extended		
	Centered	Immediate		
		30 seconds		
		60 seconds		
		90 seconds		
	Position B	Retracted		
400	Position A	Extended		
	Centered	Immediate		
		30 seconds		
		60 seconds		
		90 seconds		
	Position B	Retracted		
500	Position A	Extended		
	Centered	Immediate		
		30 seconds		
		60 seconds		
		90 seconds		
	Position B	Retracted		

Activity Analysis

Answer the following questions in relation to the collected data and observations of the test procedures.

1. Which of the circuits tested would be the most economical to operate in an actual machine? Which components in this circuit contribute to the energy savings? Explain your answer.

2. Internal component leakage is simulated in these test circuits by the needle valve. Name at least three places where wear or component damage on actual equipment could cause leakage.

3. Even if the needle valve is closed, what basic fluid concept prevents circuit 1 from maintaining pressure on the cylinder when the directional control valve is shifted to the center position?

4. Examine the gauge 1 pressures recorded in charts 1 and 2. Why are these reading similar when the direction control valve is shifted to the center position?

5. Compare the gauge 2 pressures recorded in charts 1 and 2. Why do these pressures vary between the two test circuits even though the system relief valve setting and the directional control valve center are identical?

6. Compare the gauge 2 pressures recorded in chart 2 for when the directional control valve is in the center position. Explain any differences that occur over the test periods.

7. What component and circuit operating features could be modified in circuit 2 to increase the length of time the actuator remains pressurized?

Chapter 12

Conditioning System Fluid

Filtration and Temperature Control

The activities in this lab exercise are designed to show several factors related to the design and application of fluid-conditioning components encountered in hydraulic systems. Emphasis is placed on demonstrating basic filter and heat exchanger concepts. The activities familiarize you with filter data sheets that provide information on the ability of a filter to maintain required fluid cleanliness. Information is also provided on identifying factors that need to be considered when heat must be added or removed to control fluid temperature.

Caution: Check with your instructor to be certain you are working with the correct equipment. Also, follow all safety procedures for your laboratory as you complete the activities.

Key Terms

The following terms are used in this chapter. As you read the text, record the meaning and importance of each. Additionally, you may use other sources, such as manufacturer literature, an encyclopedia, or the Internet, to obtain more information.

absolute rating ______________________________

absorbent filter ______________________________

adsorbent filter ______________________________

baffle ______________________________

bonnet ______________________________

brazed-plate heat exchanger ______________________________

bypass valve ______________________________

cleanout ______________________________

contaminant ______________________________

depth-type filter ______________________________

filter ______________________________

filter-element-condition indicator ______________________________

filtration system ______________________________

finned conductor ______________________________

fluid conditioning ______________________________

full-flow filtration ______________________________

heat exchanger ______________________________

____

____

immersion heat exchanger ______________________________

____

____

micron ______________________________

____

____

nominal rating ______________________________

____

____

off-line filtration ______________________________

____

____

proportional filtration ______________________________

____

____

radiators ______________________________

____

____

shell-and-tube heat exchanger ______________________________

____

____

sludge ______________________________

____

____

strainer ______________________________

____

____

suction filter ______________________________

____

____

sump strainer ______________________________

____

____

surface-type filter ______________________________

____

____

varnish ______________________________

____

____

Chapter 12 Quiz

Name ______________________________ Date ______________ Score ________

Write the best answer to each of the following questions in the blanks provided.

1. Name the four sources of contaminants.

____

____

____

____

2. ______ accelerates the buildup of residue on the surfaces of parts and leads to clogging of valve passageways.

____

3. Low operating temperatures cause larger-than-desired pressure drops and ______ system operation.

____

4. A basic hydraulic system component that stores fluid and is considered a basic contamination-control device is the ______.

____

5. Name the two filtration methods used by strainers and filters to remove contaminants from hydraulic fluid.

____

____

6. What is the inch equivalent of one micron?

____

7. Which filtration rating is based on an average pore size and does not guarantee removal of all larger-size contaminant particles?

____

8. A filter-element-condition indicator shows the ______ between the inlet and outlet sides of the filter.

____

9. Care must be taken to correctly size an inlet line filter as a unit with an inadequate flow capacity can cause ______, resulting in pump damage.

____

10. Name the three filtration routing methods used to direct fluid flow through system filters.

____

____

____

11. As a general rule, the output of the circulating pump used in an off-line filtration system should be ______ percent of the volume of the reservoir.

____

12. Name the two common air-cooled heat exchangers used in hydraulic systems.

____

____

13. How many times will water pass the length of the shell-and-tube heat exchanger?

____

____

____

____

14. Shell-and-tube, ______, and ______ heat exchangers have the ability to either cool or heat the hydraulic fluid passing through the unit.

____

15. Identify three factors that can be used to analyze a system for the need of a heat exchanger.

____

____

____

Activity 12-1

Identification and Analysis of Filter Specifications and Performance Information from Data Sheets

Name ______________________________ **Date** ______________

This activity is designed to show how a fluid power component manufacturer presents technical information about the fluid filters and systems it designs, manufactures, and markets. Technical data includes information relating to filter materials and level of filtration that must be considered when selecting a filter. Filter manufacturers often supply information about locating filters in the system to provide maximum filtration with minimal pressure drop across the filter element.

Activity Specifications

Analyze the data sheets for the hydraulic system filtration units assigned by your instructor. The first section of the activity requires a review of the data sheets to identify the basic information available about the filter models you have been assigned. Become familiar with the data sheets before attempting to answer activity questions. The second section of the activity involves information concerning the filter elements available for use with the filter housing.

Assigned filter

Manufacturer: ______________________________

Model designation: ______________________________

Series number: ______________________________

Section one—filter housing

1. Identify the maximum recommended operating pressure for this filter model. In which areas of the hydraulic system could this system be used? Justify your answer.

____

____

____

____

2. What is the maximum rated flow capacity of the filter housing unit?

____

3. Identify the port connection configurations available with this filter housing.

____

____

____

____

4. What method is provided by the manufacturer for mounting this housing in the hydraulic system?

____

____

____

____

5. What vertical height needs to be available in the system to mount the filter and provide adequate clearance for filter element replacement?

6. Does the housing contain a filter element bypass valve? If so, what pressure drop is required to operate the valve?

Section two—filter element

1. Describe the basic style of element used with this filter model.

2. What filtering materials are used in the construction of the elements available for use with this filter model?

3. Identify the recommended operating temperature range for the filter elements.

4. Identify the rated pressure drop across the element at the maximum recommended flow rate.

5. Discuss the effect of increased pressure drop across the filter element caused by changes in fluid temperature or viscosity and the accumulation of contaminants in the filter.

Activity 12-2

Select a Heat Exchanger for a Hydraulic System Using Specified System Parameters

Name ______________________________ **Date** ______________

If a system is operating with a fluid temperature higher or lower than recommended, maximum equipment performance and service life cannot be achieved. When extreme operating conditions are encountered, it may be necessary to incorporate a heat exchanger into a system. This activity illustrates one method that may be used to select the proper size of heat exchanger to maintain an appropriate system operating temperature.

Activity Specifications

Select an appropriate size of heat exchanger for a hydraulic system operating under the load and ambient conditions designated by your instructor. Use data sheets and recommended formulas from the heat exchanger manufacturer to determine the type and size of unit that can provide a stable operating temperature for the system. Describe the control method needed to maintain the desired system operating temperature.

Sizing a Heat Exchanger

Manufacturers often provide formulas and direct technical support for use in determining an appropriate heat exchanger for use in a hydraulic system application. These services are extremely helpful because of the design of the units and the nature of heat transfer in fluid systems. The following websites provide formulas, calculators, or engineering support that may be used in sizing exchangers for an application. Review these sites or other similar sites recommended by your instructor before continuing with this activity.

www.aihti.com

www.exergyinc.com

www.engineeringpage.com

System parameters

Obtain the following operating parameters from your instructor for a hydraulic system requiring installation of a heat exchanger.

Prime mover horsepower: ______________________________

System pump flow rate (gpm): ______________________________

System pressure (psi): ______________________________

Hydraulic system fluid: ______________________________

Pump inlet fluid temperature: ______________________________

Fluid temperature exiting the system: ______________________________

Heat exchanger fluid: ______________________________

Heat exchanger fluid flow rate (gpm): ______________________________

Heat exchanger fluid inlet temperature: ______________________________

Heat exchanger fluid outlet temperature: ______________________________

Selection procedure

Use suggested formulas and procedures from websites or your instructor to determine the model and size of heat exchanger needed to maintain a desired system operating temperature. Attach your worksheets showing the formulas applied and your calculations. Include copies of reference tables from which information was obtained during both the calculation and selection process.

> **Note:** Do *not* request engineering assistance through a website to complete this activity.

Activity Analysis

Complete the following in relation to the data collected.

1. What are the sources of the heat that must be removed to maintain a desired system operating temperature?

2. Why is it necessary to monitor the pressure on both the hot and cold sides of a heat exchanger?

3. What methods are used in the construction and operation of a heat exchanger to control the amount of heat transferred by the unit?

4. How does the specific heat of the system oil and the cooling fluid affect the performance of a heat exchanger? How are differences in specific heat compensated for in most formulas used to select a proper size of exchanger?

5. Manufacturer's selection tables often include a range of acceptable flow rates for the tube side of a heat exchanger. Provide a rationale for this flexibility.

Activity 12-3

Test the Effectiveness of a Heat Exchanger in Controlling the Operating Temperature of Hydraulic System Oil

Name ______________________________ **Date** ______________

A heat exchanger can be used in a variety of ways in a hydraulic system. In a system operating in cold weather conditions, an electric immersion heater may be used to warm the reservoir oil before the pump is started to prevent cavitation. In high ambient temperatures, an exchanger may be used to remove heat to increase the service life of system oil and prevent excessive varnish build-up on internal component parts. This activity illustrates the application of a heat exchanger in an operating power unit and how it can be used to increase or decrease reservoir oil temperature.

Activity Specifications

Construct and test a circuit to measure the performance of a shell-and-tube heat exchanger. Measure the temperature of the input and output water and the resulting temperature changes in the reservoir oil under varying circuit load conditions. Plot and compare the performance of the heat exchanger under varying water input temperatures and system loading. Use the test circuit and procedure shown below to set up and operate the circuit. Record the test data, compare the temperatures between the input flow and load conditions, and answer the activity questions.

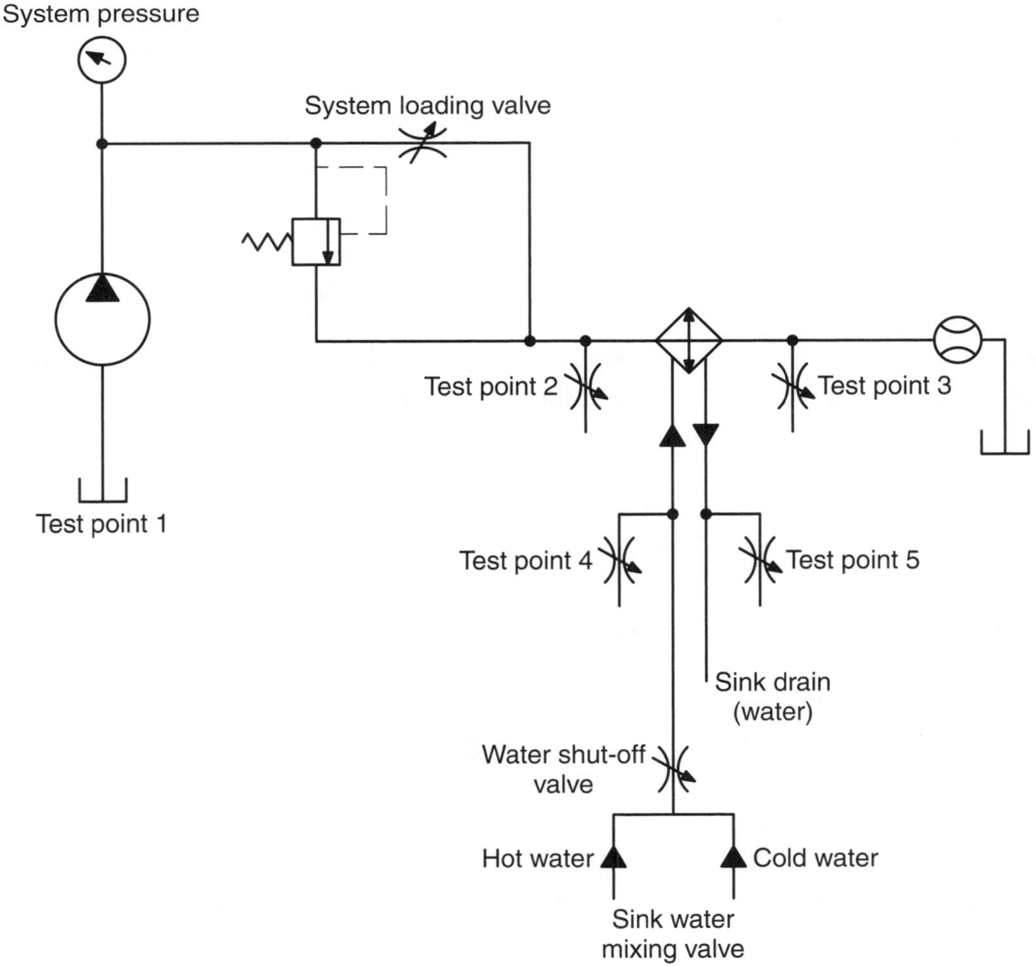

Required Components

Access to the following components is needed to complete this activity.

- Hydraulic power unit (pump, reservoir, and prime mover).
- Power unit relief valve.
- Pressure gauge (used as the power unit master pressure gauge).
- Adjustable, noncompensated flow control valve (used as the system loading device).
- Four flow control valves (small needle valves to act as the test points to obtain hydraulic fluid and heat exchanger water).
- Heat exchanger (small shell-and-tube design).
- Flowmeter (used to observe hydraulic fluid flow through the heat exchanger).
- Manifolds (adequate number to make component connections).
- Hoses (adequate number to allow assembly of the test circuit).
- Thermometer (used to measure temperature of hydraulic fluid and heat exchanger water).
- Two beakers (used to collect hydraulic fluid and water for temperature measurement).
- Utility sink with mixing valve and drain (used to control volume and temperature of water supplied to heat exchanger).
- Shut-off valve for water supply.
- Garden hoses (several sections with fittings to allow the heat exchanger to be connected to the water supply and drain).

Procedures

Study the following procedures to become familiar with the steps needed to complete the activity. Then, complete the procedures and record the observations indicated in the Data and Observation Records section.

1. Select the components needed to assemble the system and test circuit. Identify manufacturer catalog data for the heat exchanger.
2. Assemble the system. Have your instructor check your selection of components and the setup of the circuit before proceeding.
3. Identify the prime mover horsepower and the flow output of the pump. Record these in the Data and Observation Records section.
4. Calculate the volume of hydraulic fluid that will be circulated through the test circuit including the reservoir and other components. Record this in the Data and Observation Records section.
5. Close the needle valves at test points 2, 3, 4, and 5.
6. Start the power unit. Fully close the system loading valve and adjust the relief valve to operate the system at a maximum pressure of 500 psi. The cracking pressure of any safety valve built into the power unit must be set above 500 psi.
7. Stop the power unit.
8. Turn on the water supply and adjust the mixing valve to produce a water temperature of 100°F at test point 4.
9. Shut off the water supply.
10. Measure and record the initial hydraulic fluid temperature at test point 1.
11. Start the power unit.
12. Turn on the water supply and adjust the flow rate to allow a steady, but *not* turbulent flow through the exchanger.
13. Measure and record the hydraulic fluid temperature at test points 2 and 3.

14. Observe and record the hydraulic fluid flow rate through the flowmeter.
15. Measure and record the water temperature at test points 4 and 5.

> **Note:** The temperature at test point 4 should remain at 100°F throughout the total test. Readjust the water mixing valve to maintain that temperature.

16. After five minutes of system operation, measure and record the temperatures at all five test points.
17. Measure and record the temperatures after 10 minutes have elapsed and again after 15 minutes.
18. Repeat steps 7 through 17 using an input water temperature of 60°F.
19. Discuss your data and observations with your instructor before disassembling the test circuit.
20. Clean and return all components to assigned storage locations.
21. Complete the activity questions.

Data and Observation Records

Before conducting the test procedures, use manufacturer catalogs to collect published information about the heat exchanger used in the activity. Record the basic information requested for use during the analysis portion of the activity. Carefully record the temperature levels at the various test points during the test procedures.

Heat exchanger information

Basic design type: ______________________

Manufacturer: ______________________

Model number: ______________________

Fluid capacity: ______________________

Pressure rating: ______________________

General system information

Prime mover horsepower: ______________________

Pump output (gpm): ______________________

Circulated hydraulic fluid volume: ______________________

Record chart

System Water Input Temperature	Elapsed Time (minutes)	Fluid Temperatures					Flowmeter (gpm)
		Test Points					
		1	2	3	4	5	
Warm (100°F)	Initial						
	5						
	10						
	15						
Cold (60°F)	Initial						
	5						
	10						
	15						

Activity Analysis

Complete the following questions in relation to the data collected and observations of the test procedure.

1. Why is it important to estimate as accurately as possible the volume of hydraulic fluid in the system? Justify your answer.

2. Why is the fluid discharge from both the relief valve and the system loading valve routed through the heat exchanger?

3. Theoretically, how many Btus per hour will be added to the hydraulic fluid when the system is operated at the 500 psi relief valve setting without operating the heat exchanger?
   Hint: 1 hp = 2455 Btu/hr

4. Would the temperature increase at test point 1 accurately reflect the amount of heat input calculated in question 3? Justify your answer.

5. Summarize the temperature changes that occurred at test points 1 through 3 during the 15 minutes of the heat exchanger "warming test".

6. How consistent were the temperature increases during each of the 5 minute test periods summarized in question 5? Explain any pattern of variation that may have occurred.

7. Summarize the temperature changes that occurred at test points 1 through 3 during the 15 minutes of the heat exchanger "cooling test".

8. How consistent were the temperature increases during each of the 5 minute test periods summarized in question 7? Explain any pattern of variation that may have occurred.

9. Summarize the water temperature changes that occurred at test points 4 and 5 during the heat exchanger "warming" and "cooling tests".

10. How consistent was the water temperature during each of the 5 minute test periods summarized in question 9? Explain any pattern of variation that may have occurred.

11. What changes/adjustments can be made to a heat exchanger circuit, beside replacing circuit components, to increase or decrease system fluid warming or cooling?

12. Examine the flowmeter column of the chart. Explain any flow variations that appear in the meter readings. Should these flow rates be used rather than pump displacement for calculations in this activity? Justify your answer.

13. Compare the initial and 15 minute elapsed time temperatures at test point 1 of the "warming" and "cooling tests". Calculate the Btus of heat involved to produce these changes in the system hydraulic fluid.

14. How closely do the Btus calculated in question 13 match the heat input from the prime mover and the input or removal of heat by the heat exchanger? What factors contribute to the variations? Be specific in naming the contributing factors.

____

____

____

____

Chapter 13

Applying Hydraulic Power

Typical Circuits and Systems

The activities in this lab exercise are designed to show the operation of several of the basic hydraulic circuits discussed in this chapter. The activities require the assembly, operation, and analysis of selected circuits. The circuits show how a combination of basic components can achieve pressure control, speed control, and sequencing of actuators. Emphasis is placed on demonstrating those concepts key to understanding the operation of a full range of hydraulic system designs.

Caution: Check with your instructor to be certain you are working with the correct equipment. Also, follow all safety procedures for your laboratory as you complete the activities.

Key Terms

The following terms are used in this chapter. As you read the text, record the meaning and importance of each. Additionally, you may use other sources, such as manufacturer literature, an encyclopedia, or the Internet, to obtain more information.

back-pressure check valve ______________________

bleed-off circuit ______________________

deceleration circuit ______________________

deceleration valve ______________________

decompression valve ______________________

high-low pump circuit ______________________

maximum operating pressure ______________________

meter-in circuit ______________________

meter-out circuit ______________________

multiple maximum system operating pressures ______________________

overrunning load condition ______________________

pressure-reducing valve ______________________

rapid-advance-to-work circuit ______________________

reduced-pressure section ______________________

regenerative-cylinder-advance circuit ______________________

remote location ______________________

sequencing

synchronization

system cycle time

two-hand-operation circuit

vent connection

Chapter 13 Quiz

Name ________________________ Date ____________ Score ________

Write the best answer to each of the following questions in the blanks provided.

1. When maximum system pressure is reached and actuators are stalled, all pump output flow is returned to the reservoir through the ______ valve.
____
2. Name two methods that may be used to obtain multiple operating pressures in a hydraulic circuit.
____
____
3. A pressure-reducing valve must always use a(n) ______ drain.
____
4. Name the three basic flow control circuits used in hydraulic circuits.
____
____
____
5. In which flow control circuit is the flow control valve placed between the system working line and the reservoir?
____
6. The flow control circuit design that is generally considered the least accurate of the basic circuits is the ______ circuit.
____
7. The flow control circuit design that can best control an overrunning load is the ______ circuit.
____
8. Name three desirable features of rapid-advance-to-work circuits.
____
____
____
9. In the high-low pump circuit, when a high work resistance is encountered by the actuator, a(n) ______ valve isolates the discharge of the low-volume pump from the high-volume pump.
____
10. What is the basic reason for the rapid cylinder advance in a regenerative-cylinder-advance circuit?
____
____
____
____
11. Define *sequencing* in relation to hydraulic systems.
____
____
____
____

12. Connecting the power output shafts of two hydraulic motors produces an effective ______ device for use in the synchronization of actuators.

____

13. Define *freewheeling* in relation to a hydraulic motor.

____

____

____

____

14. What type of valve is a decompression valve?

____

15. Name four operating characteristics that make the two-hand-operation circuit effective as a safety device to keep both hands of the operator clear of a dangerous area.

____

____

____

____

Activity 13-1

Construct and Test a Pressure Control Circuit Providing Multiple Maximum System Operating Pressures

Name ______________________________ Date ______________

This activity is designed to show the operation of the venting feature of a system relief valve. Venting may be used to simply dump the pump output through the relief valve to quickly reduce system pressure to the lowest-possible level. It may also be used to allow a single system relief valve to provide multiple pressure levels. Venting is an option only on compound relief valves.

Activity Specifications

Construct and operate a hydraulic circuit that allows a machine operator to quickly select between three pressure levels for machine operation. Use the test circuit and procedure shown below to set up and operate the circuit. Collect the data specified in the Data and Observation Records section. Carefully analyze the circuit and the data collected. Then, answer the activity questions.

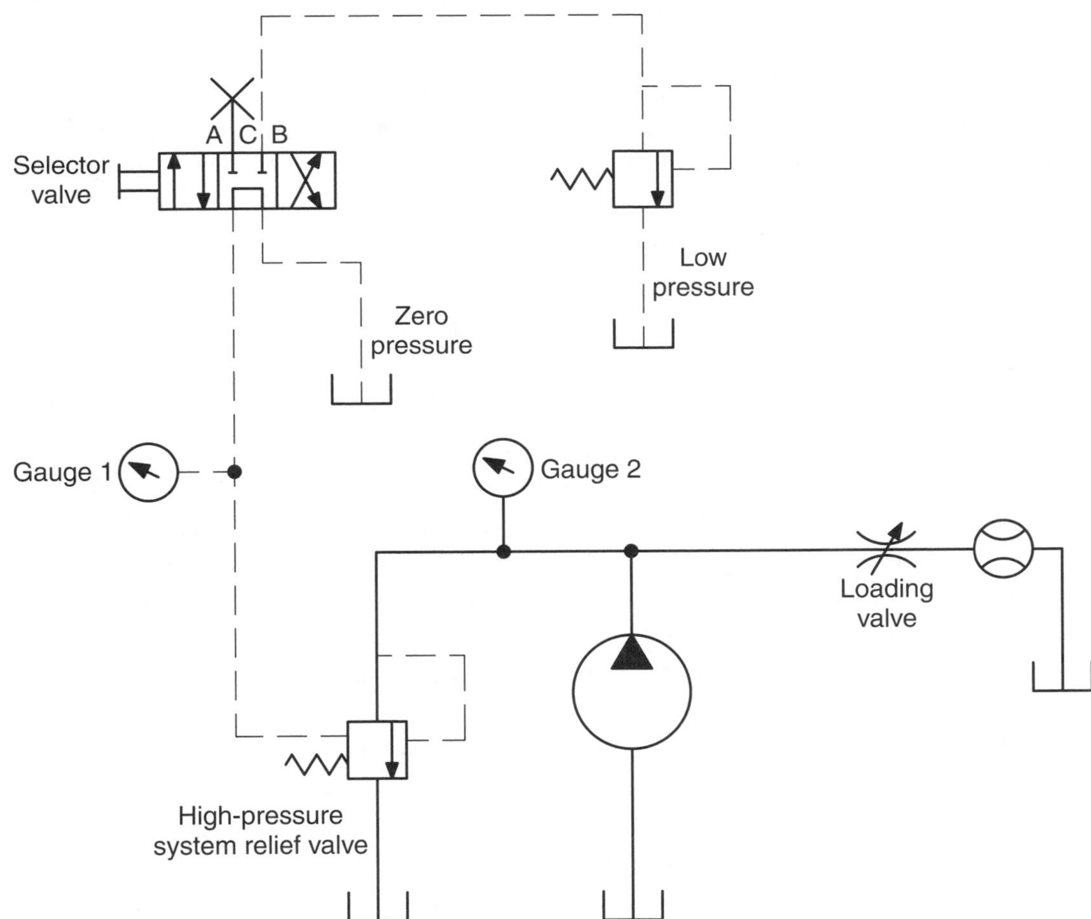

Required Components

Access to the following components is needed to complete this activity.

- Hydraulic power unit (pump, reservoir, and prime mover).
- System relief valve (compound design with control-chamber venting).
- Small, direct-operated relief valve (used to externally adjust the pressure in the control chamber of the system relief valve).
- Manually operated, three-position, four-way directional control valve with a tandem center (used as a pressure-selector valve).
- Two pressure gauges (one used to measure control chamber pressure of the system relief valve and the other to measure system operating pressure).
- Adjustable, noncompensated flow control valve (used as a loading device).
- Flowmeter (used to observe hydraulic fluid flow through loading valve).
- Manifolds (adequate number to make component connections).
- Hoses (adequate number to allow assembly of the test system).

Procedures

Study the following procedures to become familiar with the steps needed to complete the activity. Then, complete the procedures and record the observations indicated in the Data and Observation Records section.

1. Select the components needed to assemble the system and test circuit.
2. Assemble the system. Have your instructor check your selection of components and the setup of the circuit before proceeding.
3. Close the loading valve.
4. Disconnect the directional control valve from the vent port of the system relief valve.
5. Start the power unit and adjust the system relief valve to 400 psi, as indicated on gauge 2.
6. Reconnect the directional control valve to the vent port of the system relief valve.
7. Start the power unit. Shift the directional control valve to position B. Adjust the direct-operated relief valve to operate the system at a maximum pressure of 200 psi, as indicated on gauge 2.
8. Stop the power unit.
9. Completely open the system loading valve to simulate a no-load condition.
10. Shift the directional control valve to position A. Note and record in the Data and Observation Records section the pressures on gauges 1 and 2 and the fluid flow through the flowmeter.
11. Shift the directional control valve to position B. Note and record the pressures on gauges 1 and 2 and the fluid flow through the flowmeter.
12. Shift the directional control valve to position C. Note and record the pressures on gauges 1 and 2 and the fluid flow through the flowmeter.
13. Simulate a light load on the system by partly closing the loading valve.
14. Repeat steps 10 through 12. Be certain to note and record the pressures on gauges 1 and 2 and the fluid flow through the flowmeter.
15. Simulate a stalled actuator by completely closing the loading valve.
16. Repeat steps 10 through 12. Be certain to note and record the pressures on gauges 1 and 2 and the fluid flow through the flowmeter.
17. With the loading valve closed, shift the selector valve several times through positions A, B, and C. Note the time taken by the system to adjust to each of the pressure settings.
18. Discuss your data and observations with your instructor before disassembling the test circuit.

19. Clean and return all components to their assigned storage locations.
20. Complete the activity questions.

Data and Observation Records

Before conducting the laboratory tests, review the following chart to become familiar with the information that must be recorded. Be certain the data are collected under the prescribed test conditions. Carefully record pressures and flow rates at the various test points during the listed system load conditions.

System Load Condition	Pressure Selector Valve Position	Pressure		Flowmeter
		Gauge 1	Gauge 2	
No load (fully open load valve)	High; position A			
	Low; position B			
	Vented; position C			
Light load (partly open load valve)	High; position A			
	Low; position B			
	Vented; position C			
Stalled (fully closed load valve)	High; position A			
	Low; position B			
	Vented; position C			

Activity Analysis

Complete the following questions in relation to the data collected and observations of the test procedures.

1. Describe at least two advantages of using this circuit to obtain several system pressures over a design that uses multiple, "primary type" system relief valves.

2. What does gauge 2 measure? Discuss how the pressure measured affects the operation of the system relief valve.

3. Describe the varying flow rates through the flowmeter as the load increases on the system.

4. As the system load increases, which fluid lines are used to return the hydraulic fluid to the reservoir?

5. What amount of fluid is returned to the reservoir through the center position of the directional control valve and the direct-operated relief valve?

____

6. What pressure adjustment and recovery times are observed in step 17 of the test procedure? What factors in the circuit design may cause the observed times?

____

____

____

____

Activity 13-2

Construct and Test a Hydraulic Circuit Containing a Pressure-Reducing Valve

Name ______________________________ Date ______________

This activity is designed to show the operation of a pressure-reducing valve in a hydraulic circuit. These valves are used when a circuit branch requires a maximum operating pressure that is lower than the setting of the system relief valve. This type of operation requires a normally open valve that adjusts to restrict fluid flow into the branch once the desired, reduced pressure is reached.

Activity Specifications

Construct and operate a hydraulic circuit containing a pressure-reducing valve that provides a reduced pressure in a section of the circuit. Test the circuit to become familiar with the general features of the valve, including the external drain that is required for valve operation. Use the test circuit and procedure shown below to set up and operate the circuit. Collect the data specified in the Data and Observation Records section. Carefully analyze the circuit and the data collected. Then, answer the activity questions.

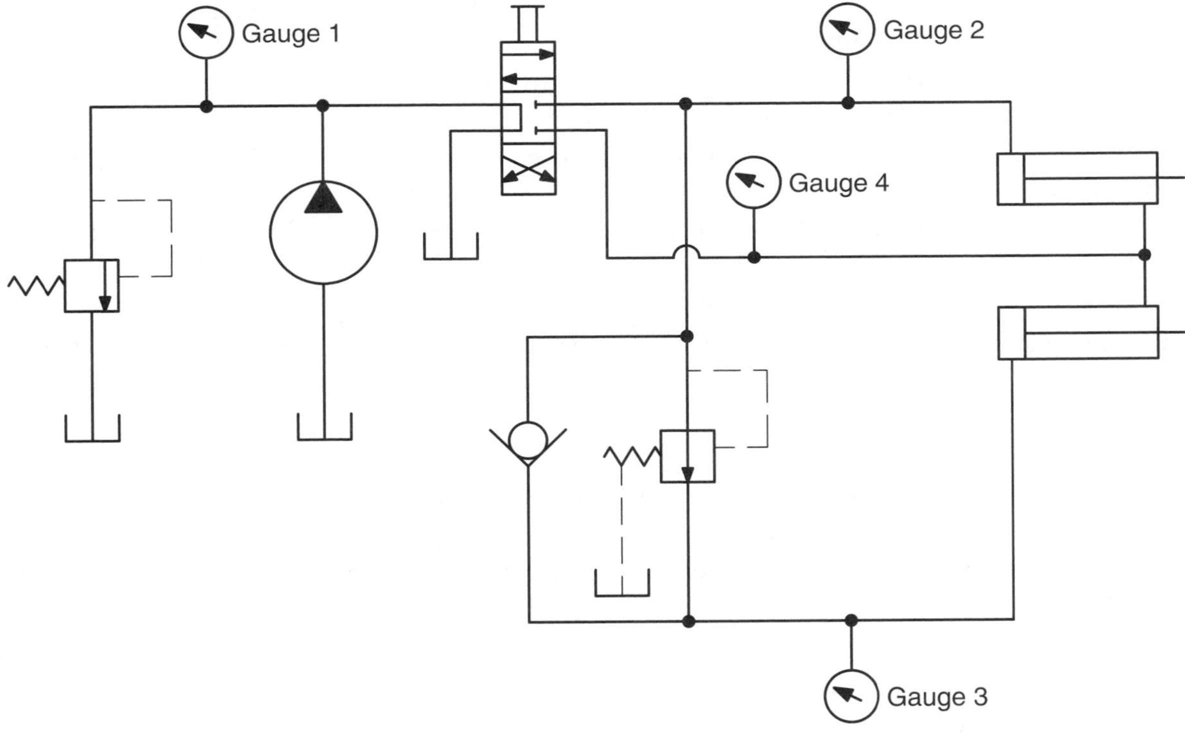

Required Components

Access to the following components is needed to complete this activity.

- Hydraulic power unit (pump, reservoir, and prime mover).
- System relief valve (compound design).
- Manually operated, three-position, four-way directional control valve with a tandem center (used to extend and retract cylinders).

- Four pressure gauges (used to test circuit pressure at various points).
- Pressure-reducing valve (used to reduce pressure in a circuit branch).
- Check valve (used to allow reverse fluid flow to bypass the pressure-reducing valve).
- Two double-acting cylinders (should be of equal size).
- Manifolds (adequate number to make component connections).
- Hoses (adequate number to allow assembly of the test system).

Procedures

Study the following procedures to become familiar with the steps needed to complete the activity. Then, complete the procedures and record the observations indicated in the Data and Observation Records section.

1. Select the components needed to assemble the system and test circuit.
2. Assemble the system. Have your instructor check your selection of components and the setup of the circuit before proceeding.
3. Start the power unit and adjust the system relief valve to a system pressure of 400 psi, as indicated on gauge 1.
4. Shift the directional control valve to fully extend both cylinders.
5. Hold the directional control in the shifted position. Adjust the pressure-reducing valve until gauge 3 reads 250 psi.
6. Shift the directional control valve to fully retract the cylinders. When the cylinders are retracted, continue to hold the valve in the shifted position. Record in the Data and Observation Records section the pressures on gauges 1, 2, 3, and 4.
7. Shift the directional control valve to extend the cylinders. While the cylinders are *moving,* note and record the pressures on gauges 1, 2, 3, and 4.
8. When the cylinders are fully extended, hold the directional valve control in the shifted position. Note and record the pressure readings on gauges 1, 2, 3, and 4.
9. Shift the directional control valve to retract the cylinders. While the cylinders are *moving,* note and record the pressure readings on gauges 1, 2, 3, and 4.
10. Repeat steps 4 through 9 with the system relief valve set at 500 psi.
11. Shift and hold the directional control valve in the extending position until both cylinders are fully extended. Quickly adjust the system relief valve setting to vary the system pressure (gauge 1) between 200 psi and 500 psi. Note any fluctuations in the gauge 3 pressure.
12. Stop the power unit.
13. Disconnect the external drain of the pressure-reducing valve.
14. Start the power unit. Shift the directional control valve to operate the cylinders through several extension and retraction cycles. Note any changes in the performance of the pressure-reducing valve and test circuit.
15. Stop the power unit.
16. Reconnect the external drain of the pressure-reducing valve.
17. Remove the check valve from the circuit.
18. Repeat step 14.
19. Discuss your data and observations with your instructor before disassembling the test circuit.
20. Clean and return all component parts to assigned storage locations.
21. Complete the activity questions.

Data and Observation Records

Before conducting the laboratory tests, review the procedures and the chart to become familiar with the information that must be recorded. Be certain the data are collected under the prescribed operating conditions.

A. Performance of the pressure-reducing valve when system pressure varies from below the set pressure of the valve to full system pressure (step 11).

B. Performance of the pressure-reducing valve and test circuit with the external drain disconnected (step 14).

C. Performance of the pressure-reducing valve and test circuit with the check valve removed (step 18).

D. Record chart

Cylinder Movement	Relief Valve	Gauge Pressure			
		1	2	3	4
Retracted	400				
	500				
Extending	400				
	500				
Extended	400				
	500				
Retracting	400				
	500				

Activity Analysis

Complete the following questions in relation to the data collected and observations of the test procedure.

1. Describe the performance of the pressure-reducing valve when the system input pressure rapidly fluctuates. What happens when system pressure drops below the pressure setting of the valve?

2. Describe the performance of the pressure-reducing valve when the drain is disconnected in step 13. Why must the pressure-reducing valve be externally drained?

3. How does system performance change when the check valve is removed from the circuit? Why?

4. Review the pressures produced when the cylinders are fully retracted and the directional control valve is held in the shifted position. Does the pressure-reducing valve appear to have an affect on the pressures? Why or why not?

5. Compare the pressures on gauges 2 and 3 when the cylinders are extending. Explain any variations in the two pressures.

6. Compare the pressures on gauges 2 and 3 when the cylinders are retracting. Explain any variation in the two pressures.

7. Compare the pressures observed on each of the system gauges during each of the system relief valve settings. Does the pressure-reducing valve appear to perform more effectively during one of these settings? Provide data and a rationale to justify your answer.

Activity 13-3

Construct and Test Basic Meter-In, Meter-Out, and Bleed-Off Flow Control Circuits

Name ______________________________ Date ______________

The movement of linear and rotary hydraulic actuators is generally associated with system fluid flow. However, this movement is influenced by both the rate of flow through the flow control valve and the location of this valve relative to the actuator. This activity illustrates the three basic circuits used with flow control valves to obtain the desired operating speed and performance of actuators.

Activity Specifications

Construct, operate, and test three basic flow control circuits. Test the circuits to become familiar with the general design of the circuits. Place special emphasis on the relative location of the flow control valve in each of the circuits and the pressures produced at the various test points. Use the test circuits and procedures shown below to set up and operate the circuits. Collect the data specified in the Data and Observation Records section. Carefully analyze the circuit and the data collected. Then, answer the activity questions.

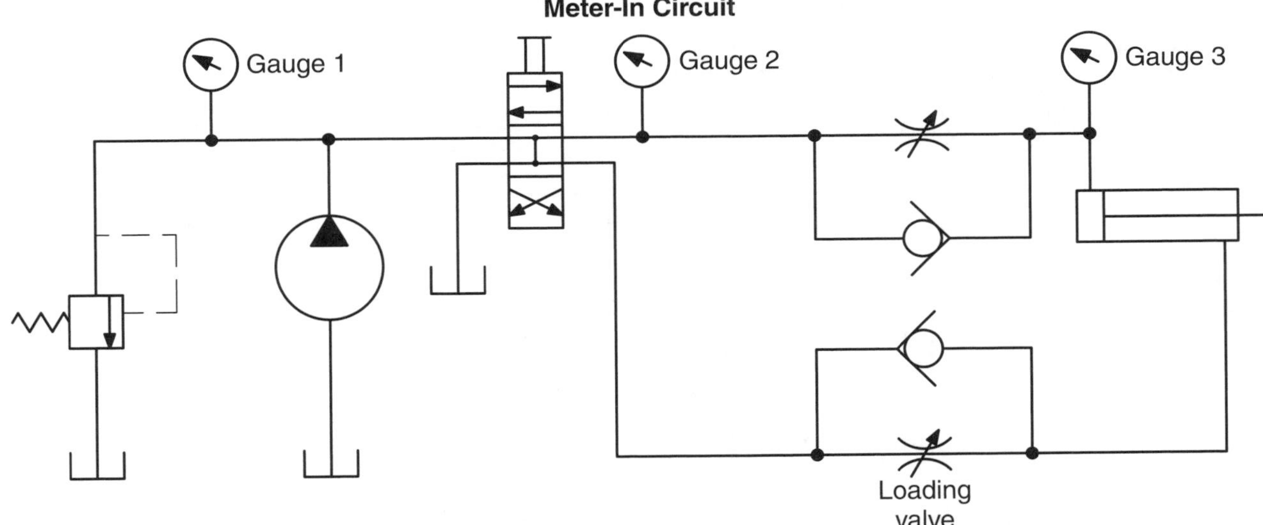

Meter-Out Circuit

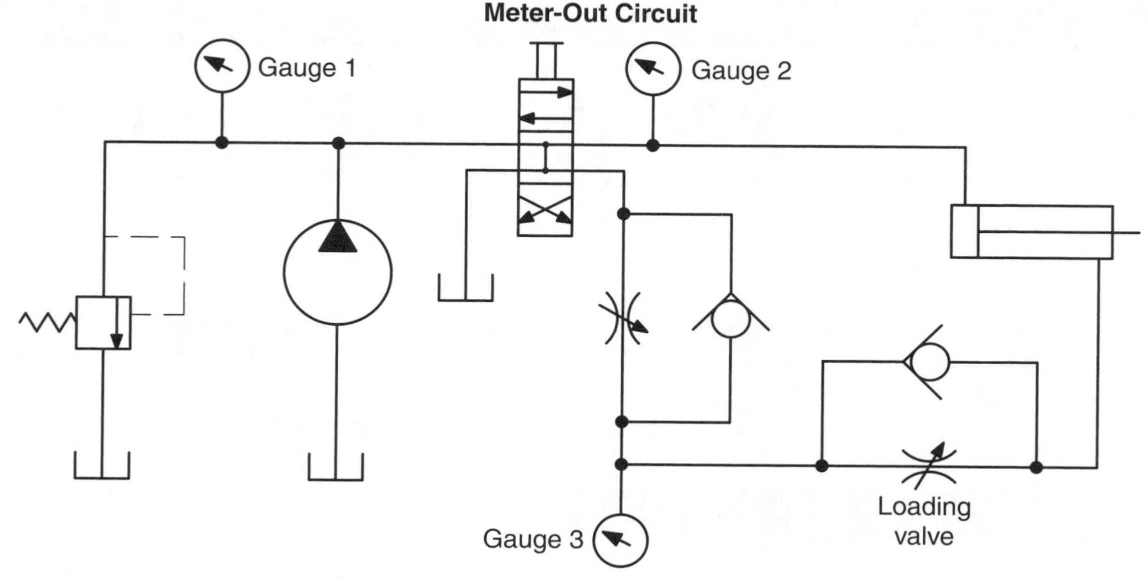

Bleed-Off Circuit

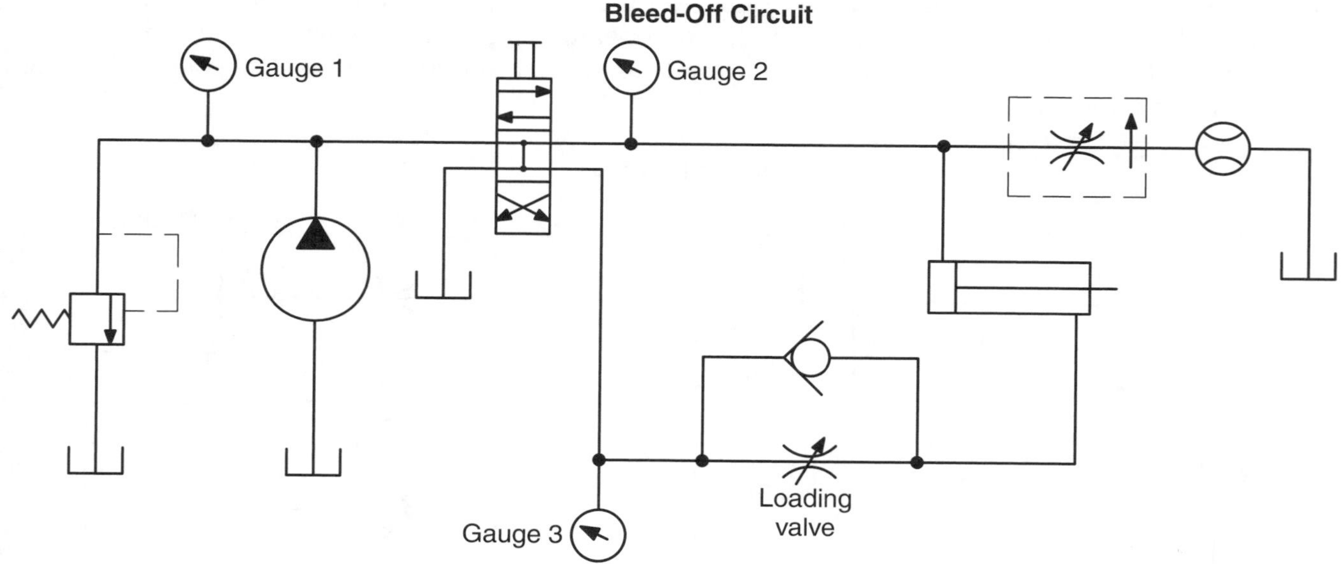

Required Components

Access to the following components is needed to complete this activity.

- Hydraulic power unit (pump, reservoir, and prime mover).
- System relief valve (compound design preferred).
- Manually operated, three-position, four-way directional control valve with a open center (used to extend and retract the cylinder).
- Three pressure gauges (used to test circuit pressure at various points).
- Two adjustable, noncompensated flow control valves with built-in check valves (one used as the flow control for actuator speed control and the other used as the actuator loading device).

- Pressure-compensated, adjustable flow control valve (used as the flow control device in the bleed-off flow control test circuit).
- Double-acting cylinder (used as the system actuator).
- Manifolds (adequate number to make component connections).
- Hoses (adequate number to allow assembly of the test system).

Procedures

Study the following procedures to become familiar with the steps needed to complete the activity. Remember, there are three separate tests using three different circuits. Then, complete the procedures and record the observations indicated in the Data and Observation Records section.

Meter-in flow control

1. Select the components needed to assemble the meter-in test circuit.
2. Assemble the system. Have your instructor check your selection of components and the setup of the meter-in test circuit before proceeding.
3. Start the power unit and adjust the system relief valve to a system pressure of 400 psi, as indicated on gauge 1.
4. Fully open the loading valve to simulate a no-load condition.
5. Adjust the flow control valve so the cylinder extends in 5 seconds.
6. Return the cylinder to the retracted position.
7. Extend the cylinder and record in the Meter-In Circuit chart the pressures on gauges 1, 2, and 3 and the extension time of the cylinder.
8. Retract the cylinder.
9. Open the loading valve one turn from the seated position to simulate a light load.
10. Extend the cylinder. Note and record gauge pressures and the cylinder extension time.
11. Retract the cylinder.
12. Open the loading valve 1/2 turn from the seated position to simulate a heavy load.
13. Extend the cylinder. Note and record gauge pressures and the cylinder extension time.
14. Retract the cylinder.
15. Fully open the loading valve to simulate a no-load condition.
16. Adjust the flow control valve so the cylinder extends in 10 seconds.
17. Repeat steps 6 through 14.
18. Stop the power unit.

Meter-out flow control

19. Relocate the flow control valve following the layout in the meter-out test circuit.
20. Start the power unit.
21. Repeat steps 4 through 18. Record in the Meter-Out Circuit chart the pressures and cylinder extension times.

Bleed-off flow control

22. Reconstruct the test circuit following the layout in the bleed-off test circuit.
23. Start the power unit.

24. Fully close the loading valve to stall the cylinder.
25. Attempt to extend the cylinder by shifting the directional control valve. While holding the directional control valve in the shifted position, adjust the bleed-off flow control valve until 2 gpm is flowing through the flowmeter.
26. Fully open the loading valve so no load is placed on the cylinder.
27. Extend the cylinder. Note and record in the Bleed-Off Circuit chart the pressures on gauges 1, 2, and 3 while the cylinder is moving, the flow through the flowmeter, and the extension time of the cylinder.
28. Open the loading valve 1 1/2 turns from the seated position to place a light load on the cylinder.
29. Repeat step 27, recording all requested data.
30. Repeat steps 28 and 29 for the following load conditions. Adjust the loading valve as indicated.
    A. Medium load (one turn from seated position).
    B. Heavy load (1/2 turn from seated position).
    C. Cylinder stalled (fully closed).
31. With the cylinder stalled during extension, adjust the bleed-off flow control until 1 gpm is returned to the reservoir through the flowmeter.
32. Repeat steps 26 through 30 for the new flow setting.
33. Stop the power unit.

Activity completion

34. Discuss your data and observations with your instructor before disassembling the test circuit.
35. Clean and return all component parts to their assigned storage locations.
36. Complete the activity questions.

Data and Observation Records

Complete the following charts based on operation of the test circuits. Be certain the data are collected following the prescribed operating conditions.

Meter-in circuit

Extension Time	Load Conditions	Pressure			Extension Time
		Gauge 1	Gauge 2	Gauge 3	
5 sec	Heavy				
	No load				
	Light				
10 sec	Heavy				
	No load				
	Light				

Meter-out circuit

Extension Time	Load Conditions	Pressure			Extension Time
		Gauge 1	Gauge 2	Gauge 3	
5 sec	Heavy				
	No load				
	Light				
10 sec	Heavy				
	No load				
	Light				

Bleed-off circuit

Bleed-Off Flow Setting	Load Condition	Pressure			Extension Time	Actual Bleed-Off
		Gauge 1	Gauge 2	Gauge 3		
2 gpm	No load					
	Light					
	Medium					
	Heavy					
	Stalled					
1 gpm	No load					
	Light					
	Medium					
	Heavy					
	Stalled					

Activity Analysis

Answer the following questions based on the pressures, flow rates, and cylinder extension times observed during the operation of the test circuits.

1. Why is the pressure reading on gauge 2 in the meter-in circuit approximately the same as the relief valve setting?

____

____

____

____

2. During meter-in circuit operation, why is the pressure on gauge 3 generally well below the relief valve setting?

____

____

____

____

3. For each of the test circuits, describe the flow path of the system oil that does not pass through the flow control valve.

4. In the meter-out circuit, explain the reason for the variations in gauge 3 pressure under the various load conditions.

5. In the meter-out circuit, what causes the pressure at gauge 3 to be higher than the relief valve setting when the cylinder is extended under a no-load condition?

6. In the meter-out circuit, what happens to the pressure on gauge 2 as the cylinder load increases? Why?

7. In both the meter-in and meter-out circuits, why does the pressure on gauge 1 not vary under loaded and no-load conditions?

8. In the bleed-off circuit, why does the pressure on gauge 1 vary with the load?

9. In the bleed-off circuit, why does the extension time of the cylinder vary when a very light or a very heavy load is placed on the system?

10. Describe the type of application each of the three flow control circuits would be best suited to handle in an actual operating hydraulic system.

____

____

____

____

11. Why are the retraction times of the cylinder uniform in each of these test circuits?

____

____

____

____

Activity 13-4

Construct and Test a Hydraulic Circuit Containing a Deceleration Valve to Control Cylinder Movement

Name ______________________________ **Date** ______________

Machines often require considerable movement to position a cutting tool or other machine member before beginning a low-speed portion of the cycle. In order to reduce the time required for this type of cycle, a hydraulic circuit must be able to provide a large volume of fluid for rapid actuator extension, followed by controlled deceleration to the beginning of the low-speed operation. This activity illustrates the operation of a circuit containing a deceleration valve that provides rapid actuator extension and deceleration followed by accurate, reduced actuator speed.

Activity Specifications

Construct, operate, and test a speed-control circuit that incorporates a deceleration valve. Use the test circuit and procedure shown below to set up and operate the circuit. Collect the data specified in the Data and Observation Records section. Carefully analyze the circuit and the data collected. Then, answer the activity questions.

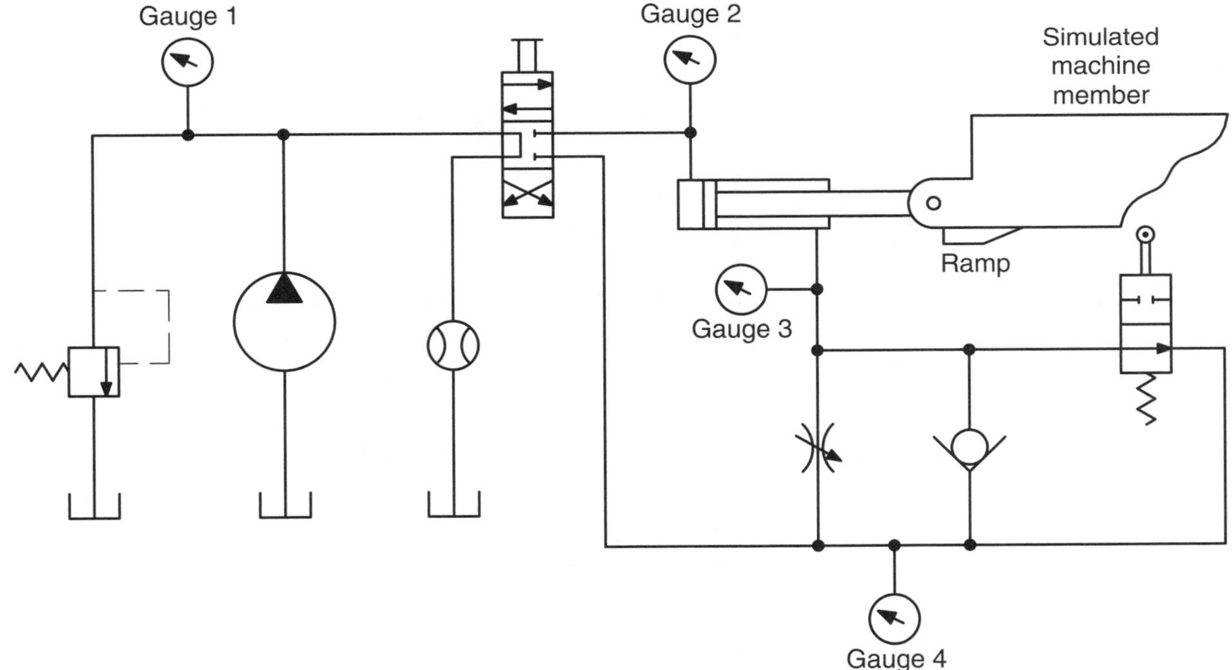

Required Components

Access to the following components is needed to complete this activity.

- Hydraulic power unit (pump, reservoir, and prime mover).
- System relief valve (compound design preferred).
- Manually operated, three-position, four-way directional control valve with tandem center (used to extend and retract the cylinder).

- Normally open deceleration valve (used to gradually reduce fluid flow from the cylinder to reduce extension speed).
- Noncompensated, adjustable flow control valve (used as the flow control device in the control test circuit).
- Double-acting cylinder (used as the system actuator).
- Machine ramp (collar that slides over the cylinder rod and serves as a ramp to mechanically operate the deceleration valve).
- Constructed deceleration valve and cylinder mounting platform (allows the ramp located on the end of the cylinder rod to depress the deceleration valve).
- Four pressure gauges (used to test circuit pressure at various points).
- Flowmeter (used to measure output flow from the cylinder during extension and retraction).
- Manifolds (adequate number to make component connections).
- Hoses (adequate number to allow assembly of the test system).

Procedures

Study the following procedures to become familiar with the steps needed to complete the activity. Then, complete the procedures and record the observations indicated in the Data and Observation Records section.

1. Select the components needed to assemble the test circuit.
2. Assemble the system. Have your instructor check your selection of components and the setup of the deceleration valve and ramp of the simulated machine member.
3. Start the power unit and adjust the relief valve to a system pressure of 400 psi, as indicated on gauge 1.
4. Shift the directional control valve to completely retract the cylinder.
5. Completely close the flow control.
6. Shift the directional control valve to extend the cylinder. Observe the operating speed of the cylinder and the flow through the flowmeter as the machine ramp depresses the deceleration valve actuator.
7. Shift the directional control valve to retract the cylinder. Observe the operating speed of the cylinder and the flow through the flowmeter as the machine ramp moves off of the deceleration valve and the cylinder continues to move to the fully retracted position.
8. Open the flow control valve 1/4 turn.
9. Shift the directional control valve to extend the cylinder. Note and record in the Data and Observation Records section the amount of cylinder travel time:
   A. From the retracted position to the point where the deceleration valve contacts the ramp.
   B. When the deceleration valve actuator is on the ramp incline.
   C. When the deceleration valve is fully depressed.
10. Shift the directional control valve to retract the cylinder. Note and record the amount of cylinder travel time:
    A. From the fully extended position to the point where the deceleration valve actuator contacts the ramp.
    B. When the deceleration valve actuator is on the ramp incline.
    C. From when the ramp moves off of the deceleration valve to the fully retracted position.

11. Repeat step 9 noting and recording the pressure on each of the four gauges at each of the three cylinder travel phases.
12. Repeat step 10 noting and recording the pressure on each of the four gauges at each of the three cylinder travel phases.
13. Repeat step 9 noting and recording the flow through the flowmeter during each of the three cylinder travel phases.
14. Repeat step 10 noting and recording the flow through the flowmeter during each of the three cylinder travel phases.
15. Open the flow control valve 1/2 turn.
16. Repeat steps 9 though 14 using the new flow control valve setting.
17. Stop the power unit.
18. Discuss your data and observations with your instructor before disassembling the test circuit.
19. Clean and return all components to their assigned storage locations.
20. Complete the activity questions.

Data and Observation Records

Before conducting the tests, review the listed procedures and the chart to become familiar with the information that must be recorded. Be certain the data are collected under the prescribed operating conditions.

A. Observed performance of the system during cylinder *extension* when the flow control valve is completely closed (step 6).

____

____

____

____

B. Observed performance of the system during cylinder *retraction* when the flow control valve is completely closed (step 7).

____

____

____

____

C. Record chart

Cylinder Travel Phase	Flow Valve Setting	Cylinder Movement	Time	Flowmeter	Pressure Gauges			
					1	2	3	4
Retracted to ramp	1/4 turn							
	1/2 turn							
Ramp	1/4 turn							
	1/2 turn							
Deceleration valve depressed	1/4 turn							
	1/2 turn							

Activity Analysis

Answer the following questions based on the pressures, flow rates, and cylinder extension times observed during the operation of the deceleration valve test circuit.

1. How would the circuit operate if the check valve is removed?

2. Which phase of circuit operation produced the greatest fluid flow through the flowmeter? Why?

3. Which phase of circuit operation produced the lowest fluid flow through the flowmeter? What route does the remainder of the fluid follow to return to the reservoir?

4. Explain the variations among the pressure gauges during cylinder extension when the deceleration valve is fully depressed.

5. Describe the flow of fluid through the circuit during the phase in which the ramp is depressing the deceleration valve.

6. Which factors control cylinder deceleration in circuits of this type? Describe factors in addition to those that are basic deceleration valve design features.

7. Describe which factors in this circuit could be changed to increase or decrease the rate of cylinder deceleration.

8. Which factors determine the rate at which the cylinder retracts in this circuit? Does the deceleration valve help determine this rate? Explain your answer.

Activity 13-5

Construct and Test a Hydraulic Circuit to Sequence Two Actuators

Name ______________________________ Date ____________

Hydraulic system sequencing involves the control of actuators to provide a specific order of movement. A basic example is a circuit in which cylinder 1 extends followed by cylinder 2 once the movement of cylinder 1 is completed. This type of motion can be achieved in hydraulic systems using pressure sensing, mechanical operation of valves, or electrical control. This activity illustrates the operation of a circuit using pressure-sensing valves to provide the progressive steps in moving actuators.

Activity Specifications

Construct and operate a hydraulic sequencing circuit providing movement control of two cylinders. The sequence of operation is:

1. Cylinder 1 extends.
2. Cylinder 2 extends.
3. Cylinder 2 retracts.
4. Cylinder 1 retracts.

Use the test circuit and procedure shown below to set up and operate the circuit. Test the circuit to become familiar with the general design. Place special attention on the pressure developed in each segment of the circuit as cylinder movement is initiated. Collect the data specified in the Data and Observation Records section. Carefully analyze the circuit and the data collected. Then, answer the activity questions.

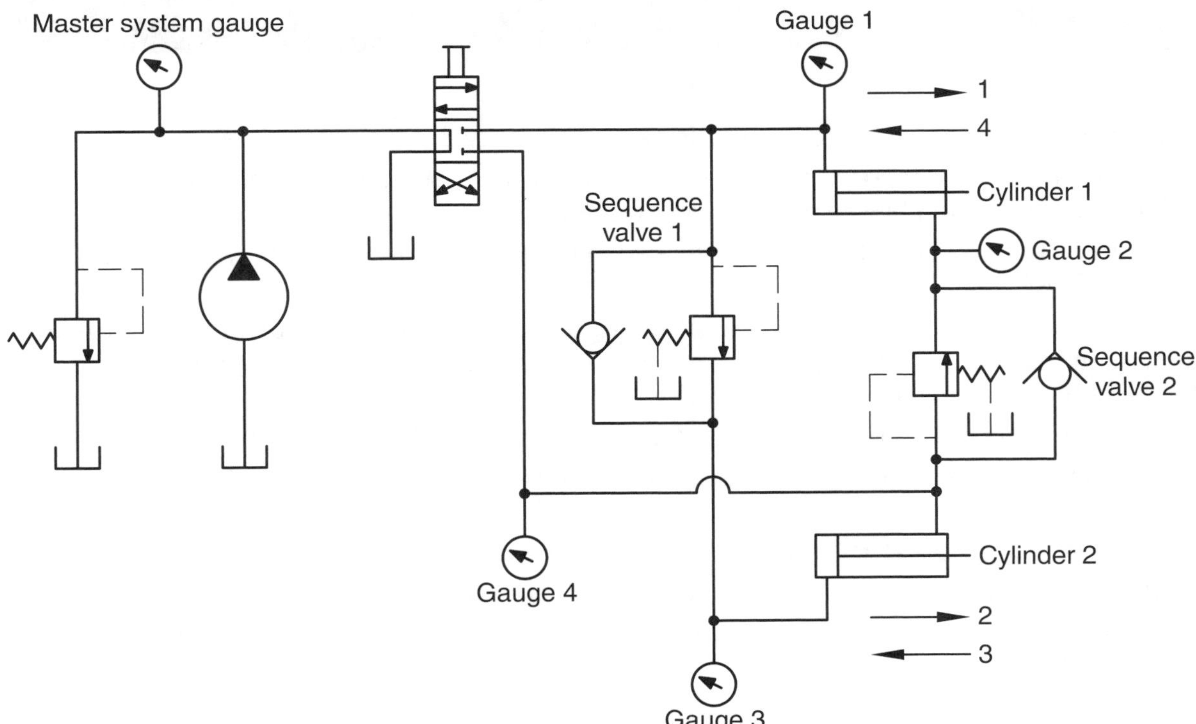

Required Components

Access to the following components is needed to complete this activity.

- Hydraulic power unit (pump, reservoir, and prime mover).
- System relief valve (compound design preferred).
- Manually operated, three-position, four-way directional control valve with tandem center (used to extend and retract cylinders).
- Two sequence valves (used to sequence cylinder movement).
- Two check valves (used to provide return flow around the sequence valves).
- Two double-acting cylinders (used as the system actuators).
- Five pressure gauges (used to test circuit pressure at various points).
- Manifolds (adequate number to make component connections).
- Hoses (adequate number to allow assembly of the test system).

Procedures

Study the following procedures to become familiar with the steps needed to complete the activity. Then, complete the procedures and record the observations indicated in the Data and Observation Records section.

1. Select the components needed to assemble the test circuit.
2. Assemble the system. Have your instructor check your selection of components and the setup of the sequence circuit.

Set sequence valve operating pressures

3. Turn the pressure adjustment screws of the sequence valves in until the adjustment spring is fully compressed.

Caution: Do *not* force the adjustment screws. Doing so can damage the valves.

4. Start the power unit.
5. Shift the directional control valve to fully extend cylinder 1.
6. While holding the directional control valve in the shifted position, adjust the system relief valve to 250 psi, as indicated by the pressure on master system gauge.
7. Continue to hold the directional control valve in the shifted position with cylinder 1 in the extended position.
8. Set the cracking pressure of sequence valve 1 to 250 psi by very slowly turning the pressure adjustment out while watching for extension of cylinder 2. The valve is properly set when cylinder 2 just begins to extend.
9. Shift the directional control valve to retract the cylinders.
10. When cylinder 2 has fully retracted, continue to hold the directional valve shifted.

Test sequence circuit operation

11. Set the cracking pressure of sequence valve 2 to 250 psi by very slowly turning the pressure adjustment out until cylinder 1 just begins to retract. The second sequence valve is properly set when the cylinder just begins to move.

12. Repeat steps 5 and 6, setting the system relief valve to 500 psi.
13. Shift the directional control valve to fully retract both cylinders.
14. Shift the directional control valve to extend the cylinders. Observe the operation of the cylinder sequencing and the pressure on the gauges as the cylinders move through the extension sequence. Make notes of your observations in the Data and Observation Records section.
15. Shift the directional control valve to retract the cylinders. Observe the operation of the cylinder sequencing and the pressures on the gauges as the cylinders move through the retraction sequence. Make notes of your observations in the Data and Observation Records section.
16. Shift the directional control valve to extend the cylinders. Note and record the pressures on gauges 1 through 4 as each cylinder moves through the extension portion of the sequence.
17. Hold the directional control valve in the extending position after both cylinders are fully extended. Note and record the pressures on gauges 1 through 4 with each cylinder in the stalled/extended position.
18. Shift the directional control valve to retract the cylinders. Note and record the pressures on gauges 1 through 4 as each cylinder moves through the retraction portion of the sequence.
19. Hold the directional control valve in the retracting position after both cylinders are fully retracted. Note and record the pressures on gauges 1 through 4 with each cylinder in the stalled/retracted position.
20. Repeat steps 5 and 6, setting the system relief valve to 400 psi.
21. Repeat steps 13 through 19 to test the operation of the circuit at the lower system pressure of 400 psi.
22. Stop the power unit.
23. Discuss your data and observations with your instructor before disassembling the test circuit.
24. Clean and return all components to their assigned storage locations.
25. Complete the activity questions.

Data and Observation Records

Before conducting the laboratory tests, review the listed procedures and the chart to become familiar with the information that must be recorded. Be certain the data are collected under the prescribed operating conditions.

A. Observed operation and performance of the sequence circuit during the extension of the cylinders. Comment on both system operating pressures (step 14).

____

____

____

____

B. Observed operation and performance of the sequence circuit during the retraction of the cylinders. Comment on both system operating pressures (step 15).

____

____

____

____

C. Record chart

System Pressure Setting	Cylinder Number	Cylinder Movement	Pressure			
			Gauge 1	Gauge 2	Gauge 3	Gauge 4
400 psi	1	Extending				
		Extended				
		Retracting				
		Retracted				
	2	Extending				
		Extended				
		Retracting				
		Retracted				
500 psi	1	Extending				
		Extended				
		Retracting				
		Retracted				
	2	Extending				
		Extended				
		Retracting				
		Retracted				

Activity Analysis

Answer the following questions based on circuit structure, system pressures, and movement involved in the test circuit.

1. Based on the test circuit, explain why sequence valves need to be externally drained.

____

____

____

____

2. What determines the maximum force that can be generated by cylinder 1 without activating cylinder 2?

____

____

____

____

3. What causes the pressure readings on gauges 2 and 4 during extension of cylinder 1?

____

____

____

____

4. Why are the pressures during cylinder retraction generally higher than during cylinder extension? Do these differences have anything to do with the operation of the basic sequence circuit?

____

____

____

____

5. Why is cracking pressure used rather than full-flow pressure when setting up a sequence valve?

____

____

____

____

6. Design a sequence circuit for a system containing a hydraulic cylinder and motor. The cylinder extends first followed by motor rotation. The motor should turn in a single direction and not rotate during cylinder retraction.

Activity 13-6

Construct and Test a Circuit Illustrating Basic Hydraulic Motor Control

Name ______________________________ **Date** ______________

The application of hydraulic motors in a circuit is generally associated with torque output and rotational speed. Although these factors are basic to circuits using motors, other features must be considered to produce additional operating characteristics. For example, it may be desirable to either control braking or allow the motor to coast to a stop. This activity illustrates basic circuit features used to obtain desired operating speed and torque, allow freewheeling, and brake without producing undesirable shock pressures.

Activity Specifications

Construct a hydraulic motor circuit allowing the motor to be powered at system pressure while incorporating features that allow the machine load to be brought to a controlled stop. Use the test circuit and procedure shown below to set up and operate the circuit. Operate and test the circuit to become familiar with the general design. Pay special attention to the valves and circuit layout that allow motor freewheeling and braking. Collect the data specified in the Data and Observation Records section. Carefully analyze the circuit and the data collected. Then, answer the activity questions.

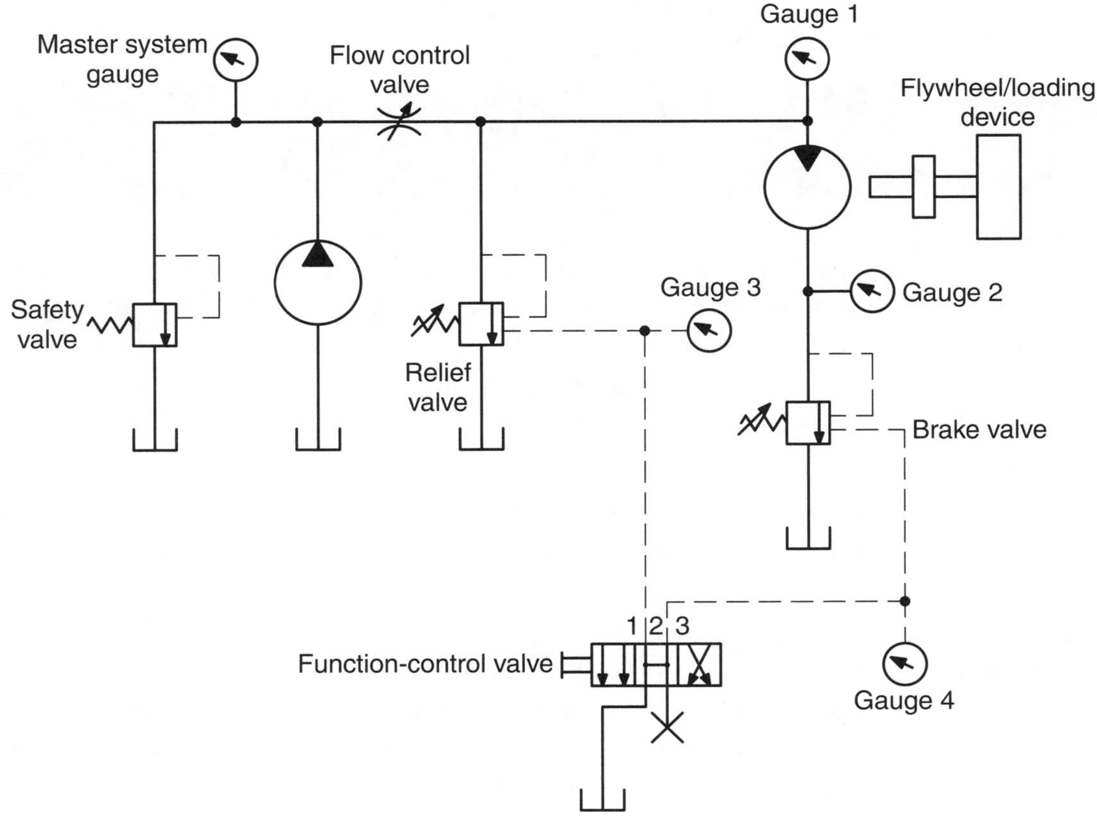

Required Components

Access to the following components is needed to complete this activity.

- Hydraulic power unit (pump, reservoir, and prime mover).
- System safety valve (compound relief valve to serve as maximum pressure valve in the circuit).
- Two relief valves (compound design with vent-control ports; one to serve as the system relief valve and the other as a motor braking valve).
- Manually operated, three-position, four-way directional control valve with open center (used as a function-control valve for the relief and brake valves).
- Flow control valve (noncompensated valve to control the speed of the hydraulic motor).
- Hydraulic motor (system actuator).
- Constructed flywheel/loading device (used to simulate motor loading and the inertia of a rotating load).
- Four pressure gauges (used to test circuit pressure at various points).
- Manifolds (adequate number to make component connections).
- Hoses (adequate number to allow assembly of the test system).

Procedures

Study the following procedures to become familiar with the steps needed to complete the activity. Then, complete the procedures and record the observations indicated in the Data and Observation Records section.

1. Select the components needed to assemble the test circuit.
2. Assemble the system. Have your instructor check your selection of components and the setup and operation of the motor flywheel/loading device.

> **Caution:** Be certain the flywheel/loading device is *securely* anchored. If not properly anchored, the torque generated by the inertia of the flywheel can cause unanticipated movement of assembly.

3. Set the maximum operating pressure of the safety valve at 500 psi, the relief valve at 300 psi, and the braking valve at 100 psi. The valves may be more easily and accurately set by removing them from the test circuit, independently setting each valve, and then replacing them in the circuit.
4. Shift the function-control valve to position 2 (centered).
5. Adjust the loading device so no load is placed on the motor.
6. Fully open the flow control valve.
7. Start the power unit.
8. Note and record the pressure readings on gauges 1 through 4 and the rotation of the motor.
9. Slowly close the flow control valve while observing any changes in gauge pressures and motor rotation.
10. Fully open the flow control valve.
11. Adjust the loading device to place a light load on the motor.
12. Note and record the pressure readings on gauges 1 through 4 and the rotation of the motor.
13. Use the loading device to increase the load on the system until the motor stalls with no indication of creeping. Note and record the pressure at each of the gauge locations.

14. Adjust the loading device to remove all load from the motor.
15. Shift the function-control valve into position 3. Hold the valve in that position while completing steps 16 through 22.
16. Note and record the pressure readings on gauges 1 through 4 and the rotation of the motor.
17. Adjust the flow control valve to vary the fluid flow rate through system. As the flow rate changes, observe any changes in motor operating speed and pressure on the four gauges.
18. Adjust the loading device to place a light load on the motor.
19. Repeat steps 16 and 17.
20. Use the loading device to increase the load until the motor stalls. Note and record the pressure on each of the gauges.
21. Adjust the loading device to remove all load from the motor.
22. Adjust the operating speed of the motor to a relatively low speed. Base this speed on the range of speeds produced as the flow control is adjusted from zero to full flow.

Caution: Check with your instructor for approval of the motor speed before proceeding to step 23.

23. Shift the function control valve into position 2 and then slowly shift it into position 1.

Caution: Carefully shift the function control valve from position 2 into position 1 until the degree of braking required by the motor speed, flywheel weight, and brake valve setting has been determined.

24. Note and record the pressure readings of gauges 1 through 4 and the rotation of the motor.
25. Shift the function control valve into position 3 to operate the motor at the previous speed setting.
26. Slightly increase the operating speed of the motor by increasing fluid flow through the flow control valve
27. Shift the function control valve into position 2 and then *slowly* into position 1.
28. Observe the pressure readings on gauges 1 through 4 and the rotation of the motor. Compare the pressures and speed with those encountered during step 24.
29. Return to step 3. Set the pressure of the brake valve to 200 psi. Verify that the safety valve remains at 500 psi and the relief valve is still set at a maximum operating pressure of 300 psi.
30. Repeat steps 4 through 28. Note and record all requested information using the 200 psi setting on the brake valve.
31. Return to step 3. Set the relief valve to 400 psi and the brake valve to 100 psi. Verify that the safety valve remains at 500 psi.
32. Repeat steps 4 through 28. Note and record all requested information using the 400 psi relief valve setting and the 100 psi brake valve setting.
33. Return to step 3. Set the brake valve to 200 psi. Verify that the safety valve remains at 500 psi and the relief valve is still operating at 400 psi.
34. Repeat steps 4 through 28. Note and record all requested information using the 200 psi setting on the brake valve.
35. Stop the power unit.
36. Discuss your data and observations with your instructor before disassembling the test circuit.
37. Clean and return all components to their assigned storage locations.
38. Complete the activity questions.

Data and Observation Records

Before conducting the laboratory tests, review the listed procedures and the chart to become familiar with the information that must be recorded. Be certain the data are collected under the prescribed operating conditions.

Valve position 1

Valve Relief Setting	Brake Valve Setting	Pressure				Motor Load	Motor Response
		Gauge 1	Gauge 2	Gauge 3	Gauge 4		
300 psi	100 psi					None	
						Light	
						Stall	
	200 psi					None	
						Light	
						Stall	
400 psi	100 psi					None	
						Light	
						Stall	
	200 psi					None	
						Light	
						Stall	

Valve position 2

Valve Relief Setting	Brake Valve Setting	Pressure				Motor Load	Motor Response
		Gauge 1	Gauge 2	Gauge 3	Gauge 4		
300 psi	100 psi					None	
						Light	
						Stall	
	200 psi					None	
						Light	
						Stall	
400 psi	100 psi					None	
						Light	
						Stall	
	200 psi					None	
						Light	
						Stall	

Valve position 3

Valve Relief Setting	Brake Valve Setting	Pressure				Motor Load	Motor Response
		Gauge 1	Gauge 2	Gauge 3	Gauge 4		
300 psi	100 psi					None	
						Light	
						Stall	
	200 psi					None	
						Light	
						Stall	
400 psi	100 psi					None	
						Light	
						Stall	
	200 psi					None	
						Light	
						Stall	

Activity Analysis

Answer the following questions based on circuit structure, system pressures, and motor operation.

1. Which function-control valve position provides:
   A. Maximum motor torque? ______________________________
   B. Motor freewheeling? ______________________________
   C. Motor braking? ______________________________
2. What aspect of circuit operation is indicated by the pressures on gauges 3 and 4?

   ____
   ____
   ____
   ____
3. Describe the operation of the circuit when the function-control valve is held in position 3.

   ____
   ____
   ____
   ____
4. Describe the operation of the circuit when the function-control valve is held in position 2.

   ____
   ____
   ____
   ____

5. What torque can be produced by the motor when the function-control valve is in position 1? Why?

6. Describe the operation of the motor when the function-control valve is shifted from position 3 to position 1. How would circuit operation change if the pressure setting of the braking valve is increased or decreased?

7. What degree of loading is necessary to stall the motor when the function control valve is in position 2? Why?

8. Explain the pressure changes that occur on gauges 1 through 4 when the function-control valve is in position 3 and the motor is varied between no load and stalled.

9. Identify all of the possible sources for the makeup fluid that is needed when the momentum of the flywheel continues to turn the motor after the function-control valve is shifted into position 2.

10. Describe features that could be incorporated into this circuit to provide a minimum delay between the shifting of the function-control valve and circuit response time.

Chapter 14

Compressed Air

The Energy Transmitting Medium

A number of relationships exist between atmospheric air and the compressed air in a pneumatic system. Although many of these factors are important to engineers who design compressors and distribution systems, they are beyond what most users need to know. The following activities are designed to demonstrate a select number of basic relationships that can be helpful to both system users and pneumatic service personnel. These factors are beyond the understanding of the sizing and operating principles of the compressor and the air distribution system. The activities consist of a number of relatively simple problems showing how the various characteristics of ambient air and system operating conditions can influence compressor and system performance. The activities apply the concepts of humidity, ambient air temperature, and how air temperature, pressure, and volume influence system performance. Emphasis is placed on demonstrating how these concepts apply to a range of pneumatic systems.

Key Terms

The following terms are used in this chapter. As you read the text, record the meaning and importance of each. Additionally, you may use other sources, such as manufacturer literature, an encyclopedia, or the Internet, to obtain more information.

adiabatic compression ______________________________

adiabatic expansion ______________________________

atmosphere ______________________________

dewpoint ______________________________

dry air ______________________________

free air ______________________________

ionosphere ______________________________

isothermal compression ______________________________

isothermal expansion ______________________________

lubricant ______________________________

mesosphere ______________________________

ozone layer ______________________________

relative humidity ______________________________

saturation ______________________________

stratosphere ________________________________

____

____

troposphere ________________________________

____

____

water vapor ________________________________

____

____

Chapter 14 Quiz

Name ______________________ Date ____________ Score ________

Write the best answer to each of the following questions in the blanks provided.

1. In which layer of the atmosphere do we live?

2. How thick is the layer in question 1?

3. Name the two components besides the gases that make up atmospheric air.

4. What causes atmospheric pressure?

5. Before atmospheric air is conditioned for use in a pneumatic system, it is typically referred to as ______ air.

6. List four factors involved in conditioning of atmospheric air.

7. Liquid water forms when air temperature drops to a level known as the ______.

8. The process that assumes a constant temperature during air compression is known as ______ compression.

9. According to the general gas law, if the volume of air decreases, the pressure of the air ______.

10. What happens if the temperature of air increases or decreases?

Activity 14-1

Relationship of Humidity to the Potential for Liquid Water in a Pneumatic System

Name ______________________________ **Date** ______________

This activity is designed to show how liquid water is formed in a pneumatic system. Free air contains varying amounts of water vapor. During air compression and distribution, this vapor can be released into the system as a liquid. Understanding the basic principles that cause this water formation is an important first step to the delivery of dry air to system workstations.

Activity Specifications

Study a pneumatic system to become familiar with the amount of water vapor that may be released into a system when compressing and distributing free air under various conditions. The calculations are based on descriptions of several atmospheric-air and pneumatic-system operating conditions. Analyze the results of the calculations and answer questions concerning methods that could be used to control the formation of liquid water in the system.

Example Problem

Calculate the volume and weight of condensed water (liquid water) that would be formed in the following circuit between the compressor and the workstation regulators under the described conditions.

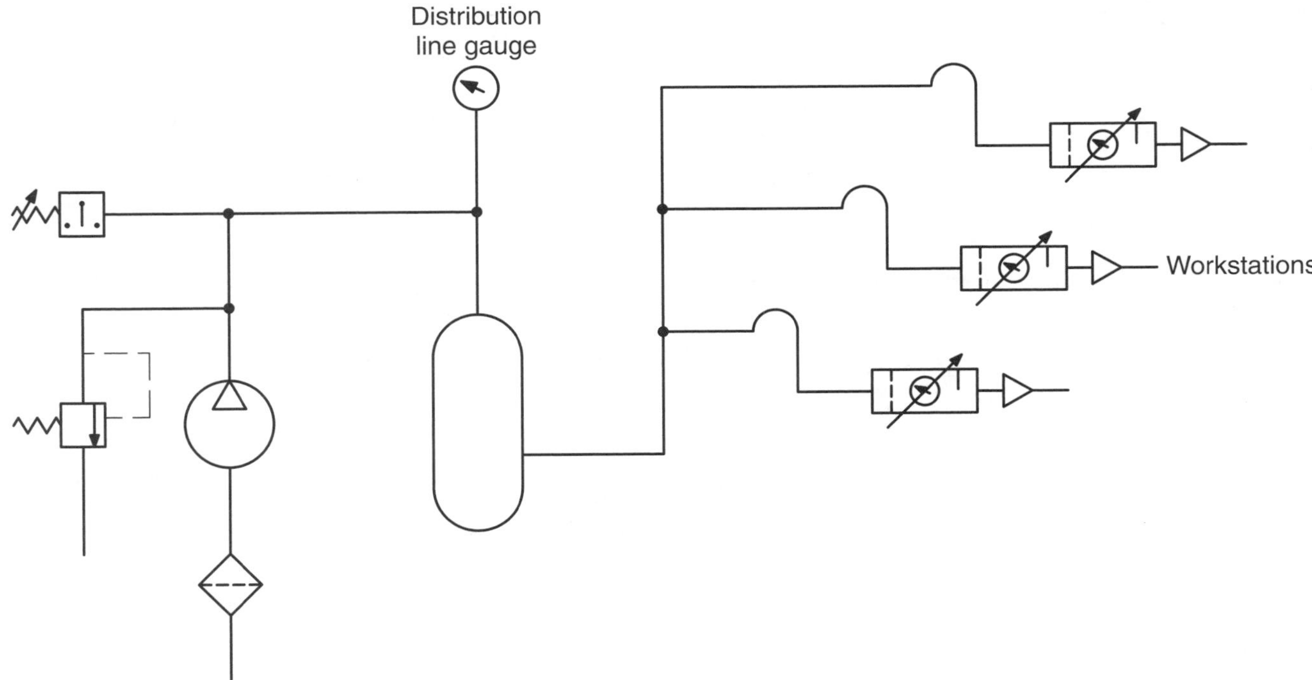

Example system air characteristics

1. Free air
   Pressure: atmospheric
   Temperature: 80°F
   Relative humidity: 60%
2. Workstation
   Air pressure: 50 psi
   Air temperature: 70°F
   Total air consumption of actuators: 40 cfm
3. Receiver and system distribution line air
   Air pressure: 80 psi
   Air temperature: 75°F

Saturated air data table

Weight of Water in 1000 Cubic Feet of Saturated Air at Various Temperatures			
Temperature (°F)	**Weight (lb)**	**Temperature (°F)**	**Weight (lb)**
–10	.040	50	.582
–5	.055	55	.693
0	.069	60	.821
5	.090	65	.980
10	.111	70	1.140
15	.144	75	1.351
20	.176	80	1.562
25	.226	85	1.837
30	.276	90	2.111
35	.338	95	2.468
40	.407	100	2.824
45	.488		

Example procedures

The following calculations can be used to approximate the weight and volume of liquid water formed during compression and distribution of air in a pneumatic system.

1. Calculate the volume of air used per hour by the workstation actuators.

   Example:

   Actuator consumption = 40 cfm

   40 cfm × 60 minutes = 2400 cubic feet per hour

2. Calculate the volume of compressed air from the distribution system consumed by the workstation per hour.

   Example:

$$\frac{P_1 \times V_1}{T_1} = \frac{P_2 \times V_2}{T_2}$$

$$\frac{80 \text{ psi} \times V_1}{\cancel{75°F}} = \frac{50 \text{ psi} \times 2400 \text{ ft}^3}{\cancel{75°F}}$$

$$V_1 = \frac{50 \cancel{psi} \times 2400 \text{ ft}^3}{80 \cancel{psi}}$$

$$= \frac{(50 + 14.7) \times 2400 \text{ ft}^3}{80 + 14.7}$$

$$= \frac{64.7 \times 2400 \text{ ft}^3}{94.7}$$

$$= \frac{155280 \text{ ft}^3}{94.7}$$

V_1 = 1639.7 ft^3 per hour

3. Calculate the volume of free air that must be compressed per hour to maintain distribution system pressure.
   Example:

$$\frac{P_1 \times V_1}{T_1} = \frac{P_2 \times V_2}{T_2}$$

$$\frac{\text{atmospheric} \times V_1}{80°F} = \frac{80 \text{ psi} \times 1639.7 \text{ ft}^3}{75°F}$$

$$V_1 = \frac{80 \cancel{psi} \times 1639.7 \text{ ft}^3 \times 80\cancel{°F}}{75\cancel{°F} \times \text{atmospheric} \cancel{psi}}$$

$$= \frac{(80 + 14.7) \times 1639.7 \text{ ft}^3 \times (80 + 460)}{(75 + 460) \times 14.7}$$

$$= \frac{94.7 \times 1639.7 \text{ ft}^3 \times 540}{535 + 14.7}$$

$$= \frac{94.7 \times 1639.7 \text{ ft}^3 \times 1.009}{14.7}$$

$$= \frac{156{,}677.1 \text{ ft}^3}{14.7}$$

V_1 = 10,658.4 ft^3 free air per hour

4. Use the Saturated Air Data Table above to determine the weight of water in the volume of free air required to power the workstation actuators. Be certain to consider the relative humidity of the free air in the calculations.

   Example:

   10,658.4 ft^3 of free atmospheric air required

   Free air temperature = 80°F

   Free air relative humidity = 60%

   Water per 1000 ft^3 of saturated air = 1.562 pounds

$$\text{weight of water in required air} = \frac{10{,}658.4\ \cancel{ft^3}\ \text{free air}}{1000\ \cancel{ft^3}} \times 1.562\ \text{pounds} \times .6\ \text{relative humidity}$$

$$= 10.6584 \times 1.562\ \text{pounds} \times .6$$

$$\text{weight of water in required air} = 9.99\ \text{pounds}$$

5. Use the Saturated Air Data Table above to calculate the weight of water in a volume of saturated free air equal to the volume of distribution line air needed to power the workstation actuators.

   Example:

   Free air temperature = 80°F

   10,658.4 ft^3 of free air needed (from step 3)

   Water per 1000 ft^3 of saturated air = 1.562 pounds

$$\text{weight of water in saturated air at 80°F} = 1.562\ \text{pounds} \times \frac{10{,}658.4\ ft^3\ \text{of saturated free air}}{1000\ ft^3}$$

$$= 16.65\ \text{pounds water}$$

6. Determine the weight of water released into the distribution system lines after the workstation by subtracting the weight of the water retained by the saturated, compressed air from the water in the incoming free air.

   Example:

   Free air temperature = 80°F

   Workstation air temperature = 70°F

   Weight of water in saturated air coming from compressor/receiver = 16.65 pounds (from step 5)

$$\text{weight of water in saturated workstation air at 70°F} = 1.140\ \text{pounds} \times \frac{10{,}658.4\ ft^3\ \text{of saturated air}}{1000\ ft^3}$$

$$= 12.15\ \text{pounds water}$$

7. Convert the weight of the water released into a volumetric measurement.

Example:

water = .0361 pounds per cubic inch

12.15 pounds = 336.6 in^3

336.6 in^3 = 1.46 gallons

1.46 gallons = 4.4 liters

Exercise Problems

Calculate the amount of liquid water that can be expected to form in pneumatic systems operating under the following conditions. Show all formulas and calculations used to determine the amount of water formed in the compressor and distribution line of each system. Refer to the previous example calculations to assist in this task.

Problem 1—large compressor

1. Free air
   Pressure: atmospheric
   Temperature: 40°F
   Relative humidity: 70%
2. Workstation
   Air pressure: 60 psi
   Air temperature: 70°F
   Total air consumption of actuators: 100 cfm
3. Receiver and system distribution line air
   Air pressure: 120 psi
   Air temperature: 80°F
4. Water in compressor: ______________________
   Water in distribution line: ______________________

Problem 2—small, portable compressor

1. Free air
   Pressure: atmospheric
   Temperature: 80°F
   Relative humidity: 90%
2. Workstation
   Air pressure: 80 psi
   Air temperature: 85°F
   Total air consumption of actuators: 3 cfm
3. Receiver
   Air pressure: 110 psi
   Air temperature: 85°F
4. Water in compressor:____________________
   Water in distribution line: ____________________

Activity Analysis

1. Define relative *humidity.* Relate relative humidity to saturated air.

2. Why must the relative humidity of the atmospheric air be considered in the calculations?

3. Which factor is the most critical to the ability of air to retain water? Why?

4. Name at least three reasons why a system designer needs to be concerned with the possibility of liquid water forming in the distribution lines of a pneumatic system.

5. What site factors (pneumatic system characteristics) might be indicators of potential water-formation problems in compressed air distribution lines.

6. Analyze the data provided for problem 2 (small, portable compressor). As you should have determined with your calculations, no liquid water will be formed if this system operates within the stated conditions. Which specific factors do you feel contribute to this?

Activity 14-2

Reaction of Air to Changes in Temperature, Pressure, and Volume

Name ______________________________ Date ______________

During pneumatic system operation, system air is constantly changing as it passes through the compressor station, distribution system, and various system actuators. During this process, the air undergoes a number of changes following the principles expressed by the basic gas laws. The many construction and operating details of a functioning pneumatic system result in constant changes in the air that make direct application of the gas law formulas impractical. However, an understanding of the principles will help in understanding the performance characteristics of both an overall pneumatic system and individual circuit segments. This activity is designed to show temperature, pressure, and volume changes that occur as elements of the gas laws change in a system.

Activity Specifications

Assemble and test simple circuits involving pneumatic components to determine volume and pressure relationships. Complete three simple tests using basic components to demonstrate how the principles of the gas laws can influence the performance of an operating pneumatic system. Operate basic start/stop compressor equipment to determine temperature changes related to air volume changes. Use the test circuits and procedures shown below to set up and operate the circuits. Analyze the information collected before answering the activity questions.

Required Components

Access to the following components is needed to complete this group of three activities.

- Small, pneumatic compressor unit (compressor, receiver, safety valve, and pressure switch).
- Workstation FRL unit (filter, regulator, lubricator combined unit).
- Double-acting cylinder (∅2″ or slightly larger with a minimum stroke of 8″; used as a test cylinder).
- Double-acting cylinder (∅1.5″ or slightly larger with a minimum stroke of 8″; used as a "pump" cylinder).
- Four needle valves (used to pressurize and vent cylinders).
- Three pressure gauges (one used to measure compressor station pressure, one to measure workstation pressure, and one to measure test cylinder pressure).
- Metal scale with a minimum division of 1/32″ or smaller (used as measuring device to determine cylinder travel).
- Electronic, high-temperature measuring device (adequate to measure temperatures over 200°F).
- Mounting platform (to hold one of the cylinders and the metal scale).
- Manifolds/connectors (adequate number to make component connections).
- Hoses (adequate number to allow assembly of the test system).

Compressor Heat Generation—Activity 14-2A

Considerable heat is generated during the operation of a compressor. The thermodynamic processes involved in the operation of any air compressor are complex. The purpose of this activity is to demonstrate the heat generated as air is compressed. The activity involves calculating the theoretical temperature increase as air is compressed during a single stroke of a compressor piston. Then, a physical measurement is made of the external surface temperature of a compressor discharge line.

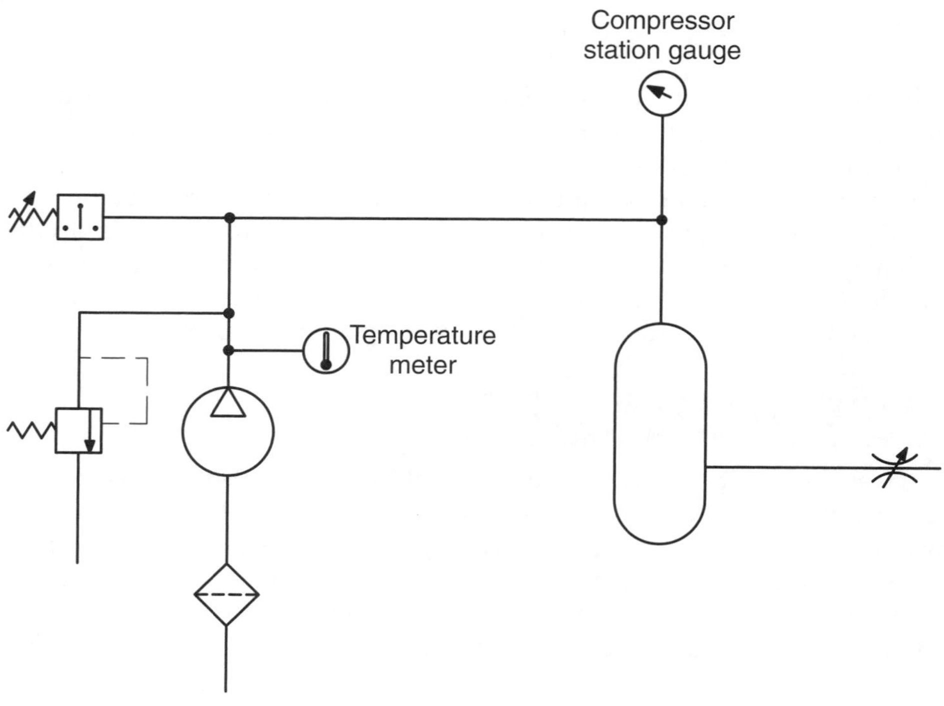

Estimate theoretical air temperature increase in compressor

Obtain the following information from your instructor for a small-displacement, single-cylinder, start/stop, portable compressor.

Cylinder bore ______________________________

Piston stroke ______________________________

Cylinder head clearance volume (estimate) ______________________________

Stop pressure setting ______________________________

Start pressure setting ______________________________

> **Note:** The following air pressure and temperature calculations are based on the general gas law. *Absolute* temperatures and pressures are required in the calculations. Show your work.

1. Calculate compressor piston displacement (displacement = area of cylinder bore × length of piston stroke).

____

2. Calculate the theoretical temperature of the air in the cylinder at the end of each compression stroke as the compressor begins to operate at the *start* pressure setting. Assume the pressure in the cylinder at the beginning of the compression stroke is 95% of the ambient air pressure.

3. Calculate the theoretic temperature of the air in the cylinder at the end of each compression stroke as the compressor operates at or near the *stop* pressure setting. Assume the pressure in the cylinder at the beginning of the compression stroke is 95% of the ambient air pressure.

Observe air temperature changes in an operating compressor

4. Identify the compressor unit and select the components needed to assemble the Activity 14-2A test circuit.
5. Attach the temperature meter sensor to the discharge line of the compressor. Place it as close as possible to the compressor outlet valve.
6. Attach the flow control valve to the outlet line of the receiver of the compressor unit. Have your instructor check your selection of components and the setup of the circuit before proceeding.
7. Note the ambient air temperature, initial temperature displayed by the meter, and pressure registered on the compressor station gauge. Record the data in the Data and Observation Records section.
8. Start the compressor and allow it to operate.
9. Note and record the outlet line temperature and the compressor station pressure when the *stop* pressure of the system is reached.
10. Partially open the flow control valve and allow air to slowly bleed from the system receiver until the compressor restarts.
11. Note and record the outlet line temperature and compressor station pressure when compressor restart occurs.
12. Operate the system through seven additional start/stop cycles. Record line temperature and system pressure for each *start* and *stop* point of the cycle.
13. Examine the data collected. Recheck any test points that appear to vary from expected pressures and temperatures.
14. Turn off the power to the compressor before continuing to Activity 14-2B.

Air Pressure Changes under Varying Volume Conditions—Activity 14-2B

This portion of the activity illustrates how the air in a stalled pneumatic cylinder will react to increasing or decreasing external loads that move the cylinder piston. The general gas law is used to predict pressure changes in the blind end of the cylinder as the piston moves. The activity simulates both increasing and decreasing loads on the cylinder. The pressure predictions are tested using the activity test circuit.

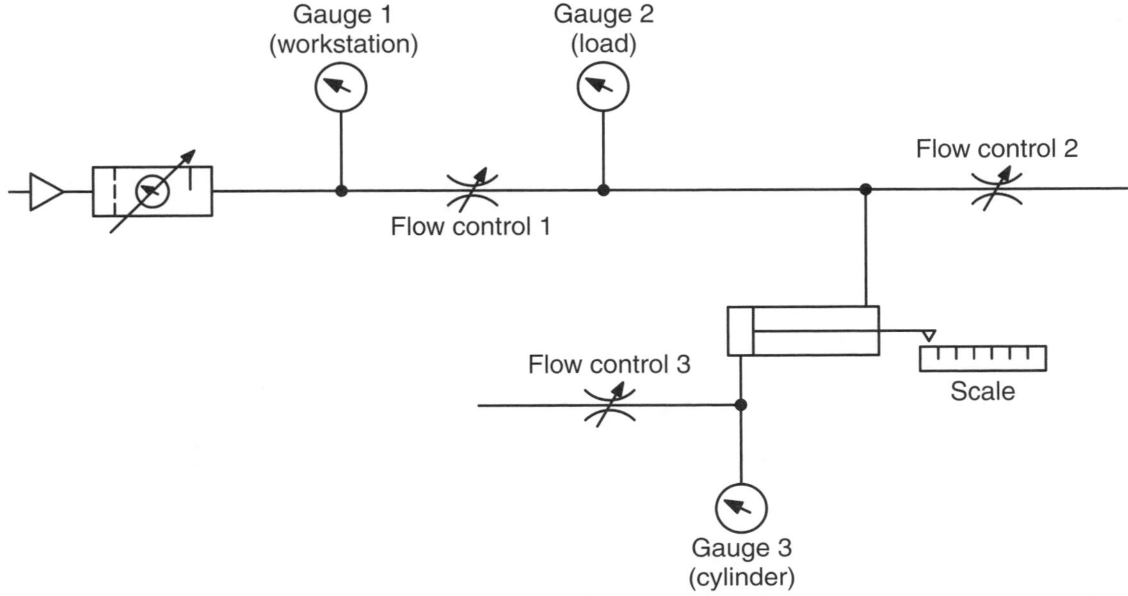

Calculate the air pressure changes that will occur in a stalled pneumatic cylinder subjected to a varying external load

1. Calculate the displacement of the blind end of the cylinder (displacement = area of cylinder bore × length of piston stroke).

2. Check the actual stroke length of the cylinder. Divide the length of the blind-end chamber into four equal-displacement sections.

3. Using the general gas law, calculate the pressure that will theoretically occur in the blind end of the cylinder as the displacement of that chamber is reduced by each one of the four displacement sections.

____

____

____

____

____

Observe changes that occur in a stalled pneumatic cylinder subjected to a varying external load

4. Select the components to assemble the Activity 14-2B test circuit.
5. Assemble the system. Have your instructor check your selection of components and the setup of the circuit before proceeding.

> **Note:** The pressure gauge and flow control valve located on the blind end of the test cylinder should be connected as closely to the cylinder fittings as possible. This will minimize the volume in the connecting lines. This volume will increase the inaccuracy of the test results.

6. Place the cylinder rod in the maximum extended position. Do this by opening flow control valves 2 and 3 and physically pulling the rod to the extended position.
7. Completely close flow control valves 1, 2, and 3.
8. Position the scale to allow an accurate reading of cylinder rod movement. Indicate on the scale each of the five displacement volumes calculated in step 2.
9. Start the compressor and allow receiver pressure to increase until the pressure switch stops the prime mover.
10. Adjust the workstation pressure regulator until gauge 1 reads 80 psi.
11. *Slowly* bleed air into the rod end of the cylinder by slightly opening flow control valve 1.
12. Close flow control 1 when the scale marker indicates the first displacement section has been compressed.
13. Record the pressures indicated on gauges 1 (workstation), 2 (load), and 3 (cylinder) in the correct chart in the Data and Observation Records section.
14. Repeat steps 11 through 13 until each of the remaining displacement sections have been compressed or until the workstation pressure setting will no longer compress the air in the cylinder.
15. Reverse the process by carefully opening flow control valve 2 to *slowly* bleed air out of the rod end of the cylinder. Be certain valve 1 is tightly closed.

16. Close flow control 2 when the scale marker indicates the first displacement section marker is reached as the cylinder rod begins to extend.
17. Record the pressures indicated on gauges 1 (workstation), 2 (load), and 3 (cylinder) in the correct chart in the Data and Observation Records section of the activity.
18. Repeat steps 16 and 17 until the cylinder is fully extended and each of the displacement sections have been checked and the pressures recorded.
19. Repeat steps 6 through 18 to collect a second set of data for each of the test points.
20. Examine the data collected and recheck any test points that appear to vary from expected pressure readings.
21. Turn off the power to the compressor before continuing to Activity 14-2C.

Predict Air Input Required to Produce a Desired Pressure under Fixed-Volume Conditions—Activity 14-2C

This portion of the activity requires application of the general gas law to predict the volume of ambient air that must be added to a fixed-volume chamber to achieve a selected system pressure. The predicted volume is then confirmed using the activity test circuit.

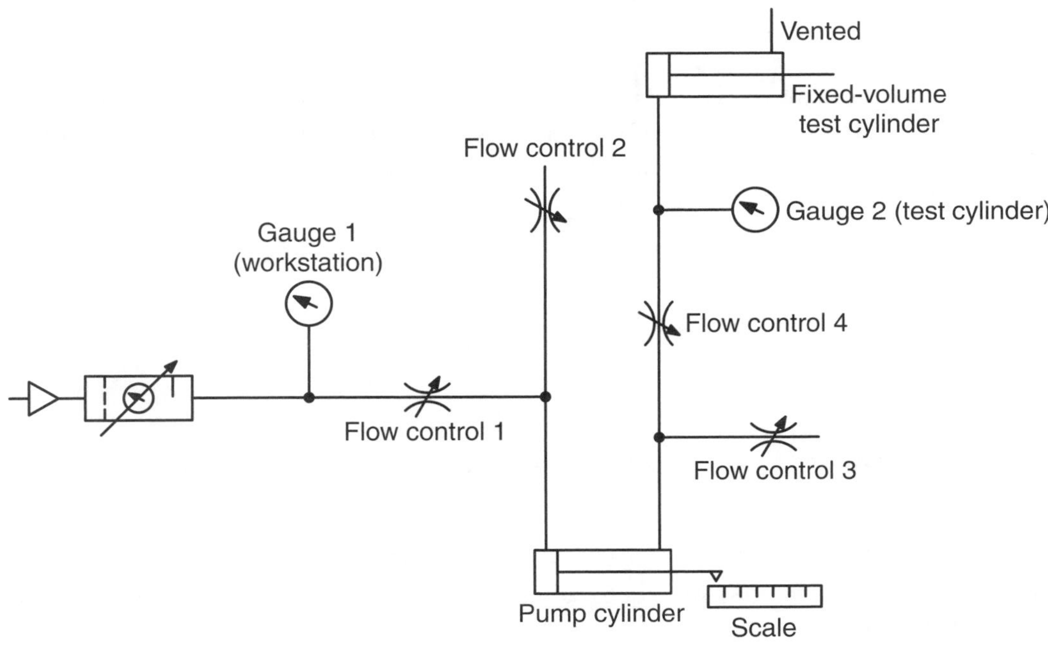

Calculate cylinder displacements and the volumes of ambient air that must be added to produce desired pressure increases

1. Calculate the displacement of the blind end of the test cylinder with the piston rod fully extended (displacement = cross-sectional area of cylinder bore × length of piston stroke).

____

2. Calculate the displacement of the rod end of the pump cylinder with the cylinder rod fully retracted (displacement = [cross-sectional area of the cylinder bore – cross-sectional area of the cylinder rod] × length of cylinder stroke).

____

3. Using the general gas law, calculate the volume of ambient air that must be added to the cylinder blind end, which serves as the fixed-volume chamber, to raise pressure to three different levels:

   40 psi ____________________ 60 psi ____________________ 80 psi ____________________

4. Using the information from the calculations in steps 2 and 3, determine the number of strokes the pump cylinder must make to move the required volume of air into the fixed-volume chamber (test cylinder) for each of the pressures.

   40 psi ____________________ 60 psi ____________________ 80 psi ____________________

Observe pressure change in a fixed-volume chamber as a measured volume of air is added

5. Select the components to assemble the Activity 14-2C test circuit.
6. Assemble the system. Have your instructor check your selection of components and the setup of the circuit before proceeding.

> **Note:** The pressure gauge and flow control valves located between the pump and test cylinders should be connected as closely to the cylinder fittings as possible. This minimizes the additional volume in the connecting lines. This volume will increase inaccuracy in the test results.

7. Place the cylinder rod of the test cylinder in the maximum extended position. Do this by opening flow control valves 3 and 4 and physically pulling the cylinder rod into the extended position.
8. Place the cylinder rod of the pump cylinder in the maximum retracted position. Do this by opening flow control valve 2 and physically pushing the cylinder rod to the retracted position.
9. Completely close flow control valves 1, 2, and 3.
10. Position the scale to allow accurate reading of cylinder rod movement.
11. Start the compressor and allow receiver pressure to increase until the pressure switch stops the prime mover.
12. Adjust the workstation pressure regulator until the workstation pressure gauge (gauge 1) reads 80 psi.
13. *Slowly* bleed air into the blind end of the pump cylinder by slightly opening flow control valve 1.

> **Caution:** Closely watch the pressure on gauge 2 to be certain the target pressure is not exceeded.

14. Close flow control valve 1 when the scale marker indicates the pump cylinder has reached the end of its stroke. Record the pressures on gauges 1 and 2 in the correct chart in the Data and Observation Records section.
15. Completely close flow control valve 4.

16. Open flow control valves 2 and 3.
17. Reposition the cylinder rod of the pump cylinder in the maximum retracted position.
18. Repeat steps 13 through 17 until gauge 2 indicates the 40 psi target pressure is reached in the fixed-volume test cylinder chamber. Close flow control valves 1 and 4 as soon as the gauge reads 40 psi.
19. Record in the Data and Observation Records section the volume of ambient air required to increase the pressure in the fixed-volume chamber to the target level.
20. Using the information calculated in steps 2 and 3, determine the number of strokes the pump cylinder must make to move the required volume of air into the fixed-volume chamber (test cylinder) to increase the pressure to 60 psi and then 80 psi.
21. Repeat steps 13 through 19 for the target pressures of 60 psi and 80 psi.

Activity completion

22. Check the charts and activity questions to be certain you have collected all of the information needed to answer the questions.
23. Discuss your observations and data with your instructor.
24. Disassemble the circuit, clean the components, and return them to their proper storage location.
25. Complete the activity questions.

Data and Observation Records

Use this section to record the results of the calculations and the observed pressure levels as each test circuit is operated. Space is provided to record information on each of the three activity sections.

Activity 14-2A

1. Calculated piston displacement of the compressor (step 1):

____

2. Theoretic temperature of the air in the cylinder at the end of the first *start* phase compression stroke (step 2):

____

3. Theoretic temperature of the air in the cylinder at the end of the last *stop* phase compression stroke (step 3):

____

4. Ambient air temperature in the vicinity of the compressor (step 7):

____

	Compressor Control Positions			
Activity Test Cycle	Restart		Stop	
	Pressure	Temperature	Pressure	Temperature
Initial				
2				
3				
4				
5				
6				
7				
8				

Activity 14-2B

1. Calculated displacement of the blind end of the cylinder (step 1):

____

2. Calculated length of the equal-displacement sections of the blind end of the cylinder (step 2):

____

3. Calculated pressures that will occur as a variable chamber is changed through four equal stages (step 3):

____

Stage 1: _______________
Stage 2: _______________
Stage 3: _______________
Stage 4: _______________

Cylinder Displacement Marker	Test Number	Cylinder Volume	Pressure		
			Gauge 1	Gauge 2	Gauge 3
1	1	Increasing			
		Decreasing			
	2	Increasing			
		Decreasing			
2	1	Increasing			
		Decreasing			
	2	Increasing			
		Decreasing			
3	1	Increasing			
		Decreasing			
	2	Increasing			
		Decreasing			
4	1	Increasing			
		Decreasing			
	2	Increasing			
		Decreasing			

Activity 14-2C

1. Calculated displacement of the blind end of the test cylinder with the piston rod fully extended (step 1):

____

2. Calculated displacement of the rod end of the pump cylinder with the cylinder rod fully retracted (step 2):

____

3. Calculated volume of ambient air that must be added to the cylinder blind end to raise the pressure in the volume chamber to (step 3):

40 psi _______________ 60 psi _______________ 80 psi _______________

4. Estimated number of pump cylinder strokes needed to reach designated test cylinder pressure (step 4):

40 psi _______________ 60 psi _______________ 80 psi _______________

Number of Pump Cylinder Strokes	Target Pressures					
	40 psi		60 psi		80 psi	
	Gauge 1	Gauge 2	Gauge 1	Gauge 2	Gauge 1	Gauge 2
1						
2						
3						
4						
5						
6						

Activity Analysis

Complete the following questions in relation to the data collected and observations made during the various test procedures.

Activity 14-2A

1. Explain the causes of the temperature variations that occur as the compressor operates through the test cycle.

____

____

____

____

2. Why do the temperatures observed at the compressor outlet vary from the calculated temperature of the compressed air?

____

____

____

____

3. Does the start/stop method of system-pressure control have an influence on the temperature of the compressed air delivered by the compressor? Explain your answer using observations made while conducting the activity.

____

____

____

____

4. What are the implications of air temperature on maintenance and efficient operation of a pneumatic system?

____

____

____

____

5. What general modes of air compression (isothermal, adiabatic, or actual) are used in this activity?

____

Activity 14-2B

1. Explain the causes of variations between actual and calculated pressure as the volume is reduced in the blind end of the cylinder.

2. Describe any variations between the pressure readings of the first and second set of collected data. Which factors contribute to these results?

3. Explain any variation between the initial pressure readings at each volume check point as the cylinder volume is decreasing and the readings taken at the same check points as air is bled out to increase volume.

4. How effective would be the long-term holding accuracy of a pneumatic cylinder? Explain your answer using data and observations from this activity.

Activity 14-2C

1. Determine the volume of air needed to raise the fixed-volume test cylinder chamber to the target pressures of 40 psi, 60 psi, and 80 psi. Base this on the number of pump cylinder strokes required to obtain the target pressure.

2. Compare the required volume to the theoretical volume calculated in step 4 of the activity procedure.

3. Why is gauge 1 pressure always lower than gauge 2 pressure in this activity? What is the basic principle involved?

Chapter 15

Source of Pneumatic Power

Compressed-Air Unit and Compressor

Compressed-air units are available in a wide variety of designs and range of capacities. The units range from small, portable compressors designed for home use to high-capacity installations involving multiple compressors supplying air to a large industrial plant. The process of selecting a compressor unit not only involves choosing a compressor design, but also establishing the required air capacity. In addition, ancillary components may be required to supply the quality of conditioned air needed to operate the equipment. The laboratory activities in this chapter are designed to illustrate data sheets supplied by compressor manufacturers and problem sheets used to estimate needed compressor capacity.

Key Terms

The following terms are used in this chapter. As you read the text, record the meaning and importance of each. Additionally, you may use other sources, such as manufacturer literature, an encyclopedia, or the Internet, to obtain more information.

air filter ______________________________

capacity-limiting system ______________________________

central air supply ______________________________

compressed-air unit ______________________________

compressor ______________________________

compressor-capacity control ______________________________

coolers ______________________________

coupling ______________________________

displacement ______________________________

double-acting compressors ______________________________

dryers ______________________________

dynamic compressors ______________________________

lobe-type compressors ______________________________

non-positive-displacement ______________________________

portable unit ______________________________

positive displacement ______________________________

prime mover ___

pumping chamber ___

pumping motion ___

receiver ___

reciprocating compressor ___

rotary compressor ___

rotary screw compressors ___

rotary, sliding-vane compressors ___

safety valve ___

single-acting compressors ___

slip ___

staging ___

Chapter 15 Quiz

Name ______________________________ Date ______________ Score ________

Write the best answer to each of the following questions in the blanks provided.

1. Name the four basic functions of compressed-air units.
____
____
____
____

2. Name the three steps involved in the operation of any basic air compressor, regardless of design.
____
____
____

3. When a compressor increases air pressure by mechanically reducing a volume of air in a compression chamber, it is a(n) ______-displacement compressor.
____

4. The compression of air by the movement of a piston in a cylinder identifies a(n) ______ compressor.
____

5. Most reciprocating compressors use pressure-operated ______ valves in the inlet and outlet ports.
____

6. The ______ reciprocating compressor design has a compression chamber located at each end of the piston.
____

7. To assure a positive air seal during compressor operation, ______ force holds the vanes of a rotary, sliding-vane compressor to the walls of the compression chamber.
____

8. Dynamic compressors use the ______ compressor design to achieve the high-speed operation needed to create airflow through the units.
____

9. The basic operating theory used by dynamic compressors is the conversion of ______ energy found in air moving at high speed to system air pressure.
____

10. In an axial-flow compressor, how is pressure created in the stator section?
____
____
____
____

11. Define *staging.*
____
____
____
____

12. Compressed air is typically discharged from multistage compressors at a(n) ______ below that produced by a comparable-sized, single-stage compressor.

____

13. The start-stop capacity-control system on small, electric motor–driven compressor units consists of a(n) ______ switch that controls the operation of the motor.

____

14. Varying the size of the compressor inlet is a capacity-control method primarily used with ______ compressors.

____

15. List the seven steps for selecting a compressor package.

____

____

____

____

____

____

____

Activity 15-1

Analysis of Information Sheets for a Small, Portable Compressor

Name ________________________________ **Date** ______________

This activity is designed to show how a manufacturer of small, portable air compressors presents technical information about the compressors it designs, manufactures, and markets. Most manufacturers supply considerable information about the compressors, although the exact content and presentation varies between companies. The approach may also vary based on whether the target market is general consumer or contractor.

Activity Specifications

Analyze and compare the data sheets of two small, portable compressors designated by your instructor. The data may be available in printed catalog sheets or on the manufacturer's website. One of the compressors should be targeted at the general consumer market. The second unit should be designed for industrial or trade-related activity. The compressors should have comparable flow capacities and pressure ratings. Carefully study the data sheets and then answer the activity questions.

Analyzed Compressors

	Consumer market	Industrial/trades market
Manufacturer		
Model		

Activity Questions

1. Identify the type and horsepower rating of the prime mover driving the compressor.

	Prime mover type	Horsepower rating
Consumer model		
Industrial/trades model		

2. Identify the design used for the construction of the compressors.
   Consumer model ________________________________
   Industrial/trades model ________________________________
3. Identify the pressure ratings of the compressors.
   Consumer model ________________________________
   Industrial/trades model ________________________________
4. Identify the airflow capacity of the compressors.
   Consumer model ________________________________
   Industrial/trades model ________________________________

5. Which methods of capacity control are used with these compressors? Why do you think the manufacturer selected these control methods?

6. Describe the path of airflow through the compressor. Concentrate on the elements that result in pressurized air for use in the system.

7. Describe the components, in addition to the prime mover and compressor, that are included in these units.

8. Which design elements justify calling these compressor packages portable?

9. List at least three consumer and three industrial/trades applications for compressor packages of these types.

Activity 15-2

Analysis of the Specification Sheets for a Portable, High-Capacity Compressor Package

Name ______________________________ **Date** ______________

Many manufacturers provide a line of compressor packages that can be transported from one work site to another. This activity is designed to show what technical information is available for those units. Although considerable information is usually provided, the exact content and presentation varies, depending on the basic design of the unit and target market. This information may be available in printed catalog sheets or on the manufacturer's website.

Activity Specifications

Analyze the data sheets of a large-capacity, portable air compressor package designated by your instructor. The package should be a skid-, trailer-, or truck-mounted unit designed for industrial applications. Carefully study the data sheets and then answer the activity questions.

Analyzed Compressor Package

Manufacturer ______________________________

Model ______________________________

Activity Questions

1. What is the physical size of this package? What means is provided to move the unit from one location to another? Are any special procedures required to assure safe transport of the unit between locations?

____

____

____

____

2. What type of prime mover is used to power the compressor? Identify the horsepower rating of the prime mover.

____

3. Identify the basic design used for the construction of the compressor. What other basic compressor designs are used by other manufacturers for this size of compressor package?

____

____

____

____

4. Identify the operating pressure range of this package. List other pressure ranges that may be available for similar high-capacity, portable units.

5. Identify the rated airflow output of the compressor in this package.

6. Describe the capacity control used with this package. Why do you think the manufacturer selected this control method for the compressor?

7. Describe the filtration used for cleaning the compressor inlet air. Why should air filtration be an especially important issue with this type compressor package?

8. Describe any design variations, other than physical size and capacity, that exist between small, portable units and this high-capacity unit.

9. List at least three industrial/commercial applications for compressor packages of this type.

Activity 15-3

Analysis of the Specification Sheets for a Large Compressor Package

Name ______________________ **Date** ______________

This activity is designed to show how the manufacturer of high-capacity compressor packages presents technical information about the units. Most manufacturers supply considerable information about these packages, although the exact content and presentation varies, depending on the basic design of the unit and target market. This information may be available in printed catalog sheets or on the manufacturer's website.

Activity Specifications

Analyze the data sheets of a high-capacity air compressor package designated by your instructor. The package should be designed primarily for stationary, industrial installations. Carefully study the data sheets and then answer the activity questions.

Analyzed Compressor Package

Manufacturer ______________________

Model ______________________

Activity Questions

1. Identify the type of prime mover driving the compressor and its horsepower rating.
2. Identify the design used for the construction of the compressor. What other basic construction types are used by other manufacturers for this size of compressor package?
3. Identify the operating pressure range of this package. List other pressure ranges that may be available in similar units.
4. Identify the rated airflow output of the compressor in this package.
5. Describe the capacity control used with this package. Why do you think the manufacturer selected this control method for the compressor?

6. Describe the components, in addition to the prime mover and compressor, that are included in this package.

____

____

____

____

7. Describe any design variations, other than physical size and capacity, that exist between portable units and this stationary compressor package.

____

____

____

____

8. List at least three industrial/commercial applications for compressor packages of this type.

____

____

____

Activity 15-4

Start-Stop Compressor-Capacity Control

Name ______________________________ Date ______________

Achieving maximum operating efficiency in a pneumatic system requires a means to closely match the output of the compressor to the air consumed by system actuators. A number of different capacity-control methods are used with pneumatic compressors to achieve this. The control method used depends on the compressor design, compressor capacity, and system application. This activity is designed to illustrate the control method commonly used with small capacity systems.

Activity Specifications

Operate a low-capacity portable air compressor equipped with start-stop capacity control to test the operation and performance of the control system. Use the test circuit and procedures shown below to set up and operate the circuit. Time the operating cycle of the compressor using three air-consumption rates. Plot and compare the operating times of the compressor and the air pressures available to the system under the three conditions. Carefully analyze the data collected and then answer the activity questions.

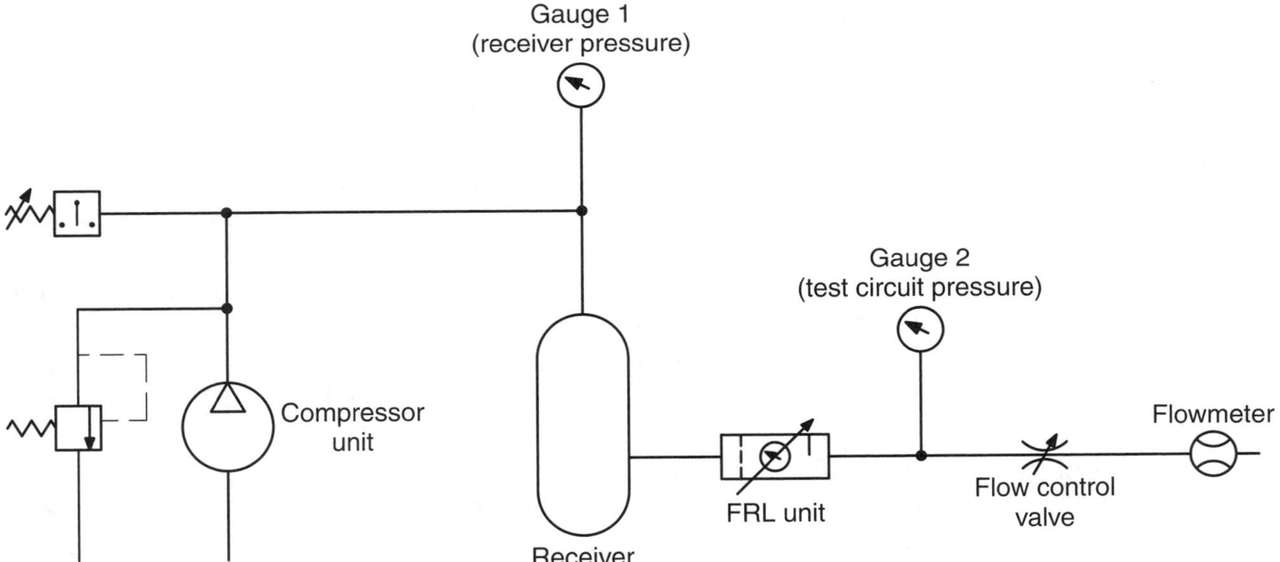

Required Components

Access to the following components is needed to complete this activity.

- Small pneumatic compressor unit (compressor, receiver, safety valve, and pressure switch).
- Workstation FRL unit (filter, regulator, lubricator combined unit).
- Adjustable flow control valve (used to adjust the rate of air exhausted to the atmosphere).
- Two pressure gauges (one used to measure compressor-station pressure and the second to measure workstation pressure).
- Timer with increments in seconds (to time the start-stop operation).

- Flowmeter (used to measure the flow rate of air exhausted to the atmosphere).
- Manifolds/connectors (adequate number to make component connections).
- Hoses (adequate number to allow assembly of the test system).

Procedures

Study the following procedures to become familiar with the steps needed to complete the activity. Then, complete the procedures and record the observations indicated in the Data and Observation Records section.

1. Select the components needed to assemble the test circuit.
2. Assemble the system. Have your instructor check your selection of components and the setup of the circuit before proceeding.

Establishing air pressure and flow characteristics for the test circuit

3. Close the flow control valve to prevent air from being exhausted to the atmosphere.
4. Adjust the pressure regulator in the FRL unit so no pressure is applied downstream of the unit when the compressor is started.
5. Start the compressor and allow receiver pressure to increase until the pressure switch stops the prime mover.
6. Note and record in the Data and Observation Records section the pressures on gauges 1 and 2.
7. Adjust the pressure regulator in the FRL unit until gauge 2 reads 50 psi.
8. Open the flow control valve 1/2 turn from the seated position. Observe the rate of air flow through the flowmeter.
9. Note the time from the opening of the flow control valve to the startup of the compressor.
10. Close the flow control valve and allow the compressor to return the receiver to full system pressure.
11. Adjust the flow control valve so the time from full system pressure to compressor restart is two minutes. This is accomplished by repeating steps 8 and 9. Slightly open or close the flow control valve to adjust the time. Several cycles may be required.
12. Record in the Data and Observation Records section the setting of the flow control valve and the rate of airflow through the flowmeter. These are the initial setting data.
13. Calculate the rate of airflow for test 2 by multiplying the initial flow rate by 1.5. Obtain the airflow for test 3 by multiplying the initial flow rate by 2.0. Record these in the Data and Observation Records section.
14. Once the opening of the flow control valve and the airflow rate through the flowmeter have been recorded, close the flow control valve and allow the compressor to bring the system to full pressure.

> **Note:** Steps 8 through 13 establish airflow test rates for the activity. The objective is to identify three graduated flow rates. Because the component parts selected for this activity vary, depending on the available equipment, it is necessary to modify the openings of the flow control valve to obtain comparable flowmeter readings.

Time the start-stop operation of the prime mover and compressor

15. Quickly open the flow control valve to obtain the initial flow rate identified in step 12.
16. Note and record the airflow rate through the flowmeter and the pressure on gauges 1 and 2 at 15 second intervals, as indicated in Chart 1 in the Data and Observation Records section.

17. Note and record the exact time the compressor restarts in the test cycle. Note this in the compressor operation column of the chart.
18. Continue to record the airflow and the pressure on gauges 1 and 2 at 15 second intervals until the compressor stops.
19. Note and record the exact time the compressor stops in the test cycle. Note this in the compressor operation column of the chart.
20. Quickly close the flow control valve to stop the loss of air from the receiver.
21. With the compressor stopped, check to be certain the receiver is fully charged to the pressure setting of the pressure switch.
22. Repeat steps 14 through 21 for airflow settings for tests 2 and 3. Record the requested flow and pressure information in charts 2 and 3.

Check the repeatability of the start-restart operation

23. Completely open the flow control valve and allow the receiver pressure to drop below the restart pressure.
24. When the compressor restarts, completely close the flow control valve so no air is exhausted.
25. When the receiver is fully recharged, note and record in the Data and Observation Records section the pressure on gauges 1 and 2. Record this information following chart 3.
26. Open the flow control valve to obtain the initial airflow setting through the flowmeter.
27. Using the timer, establish the amount of time required to drop the receiver pressure to restart the compressor.
28. Note any changes in the airflow rate through the flowmeter as the pressure drops in the compressor/receiver section of the test circuit.
29. When the compressor restarts, note the elapsed time and the air pressure on gauges 1 and 2. Record this information in the Data and Observation Records section following chart 3.
30. Establish the amount of time required to increase the receiver pressure to operate the pressure switch and stop the compressor.
31. Note the airflow rate through the flowmeter as the pressure increases in the compressor/receiver section of the test circuit.
32. Note the time required to recharge the receiver and the pressure on gauges 1 and 2. Record this information in the Data and Observation Records section following chart 3.
33. Repeat steps 23 through 32 for the airflow rate used for test 2. Record this information in the Data and Observation Records section following chart 3.
34. Repeat steps 23 through 32 for the airflow rate used for test 3. Record this information in the Data and Observation Records section following chart 3.

Start-stop capacity control when demand exceeds compressor capacity

35. Completely open the flow control valve to allow maximum airflow through the test circuit.
36. Note the operation of the test circuit including airflow rates, compressor/receiver pressures, test circuit pressures, and the general functioning of the start-stop feature of the unit.

Activity completion

37. Discuss your data and observations with your instructor before disassembling the test circuit.
38. Clean and return all components to their assigned storage locations.
39. Complete the activity questions.

Data and Observation Records

Record airflow rates, pressure readings, times, and general observations concerning the start-stop capacity control of the compressor package.

Fully pressurized system (compressor stopped, step 6):

Gauge 1 pressure ______________________________

Gauge 2 pressure ______________________________

Airflow through flowmeter (test circuit, 50 psi setting, step 12):

Flow control valve setting	Airflow
Initial	
Test 2	
Test 3	

Pressure Bleed Down to Compressor Startup

Chart 1—flow control at initial opening					
Time (min/sec)	Airflow rate	Pressure		Compressor operation	
		Gauge 1	Gauge 2	Off	Running
0:00					
0:15					
0:30					
0:45					
1:00					
1:15					
1:30					
1:45					
2:00					
2:15					
2:30					
2:45					
3:00					
3:15					
3:30					
3:45					
4:00					

Chart 2—flow control at test 2 setting					
Time (min/sec)	Airflow rate	Pressure		Compressor operation	
		Gauge 1	Gauge 2	Off	Running
0:00					
0:15					
0:30					
0:45					
1:00					
1:15					
1:30					
1:45					
2:00					
2:15					
2:30					
2:45					
3:00					
3:15					
3:30					
3:45					
4:00					

Chart 3—flow control at test 3 setting					
Time (min/sec)	Airflow rate	Pressure		Compressor operation	
		Gauge 1	Gauge 2	Off	Running
0:00					
0:15					
0:30					
0:45					
1:00					
1:15					
1:30					
1:45					
2:00					
2:15					
2:30					
2:45					
3:00					
3:15					
3:30					
3:45					
4:00					

Compressor start-stop repeatability observations

Fully pressurized system (compressor stopped)	
	Pressure
Gauge 1	
Gauge 2	

Elapsed time between full system pressure and compressor restart	
Flow control valve setting	Time (min/sec)
Initial	
Test 2	
Test 3	

Compressor restart pressure at tested workstation airflow rates	
Flow control valve setting	Time (min/sec)
Initial	
Test 2	
Test 3	

Elapsed time between compressor restart and compressor stop	
Flow control valve setting	Time (min/sec)
Initial	
Test 2	
Test 3	

Observations made during steps 35 and 36

Activity Analysis

Complete the following plots and answer the questions in relation to the data collected and observations of the test procedures.

1. Plot the data recorded for each of the airflow rates. Mark the start and stop pressure points for each of the rates. This graph will be used to support answers to the remaining questions.

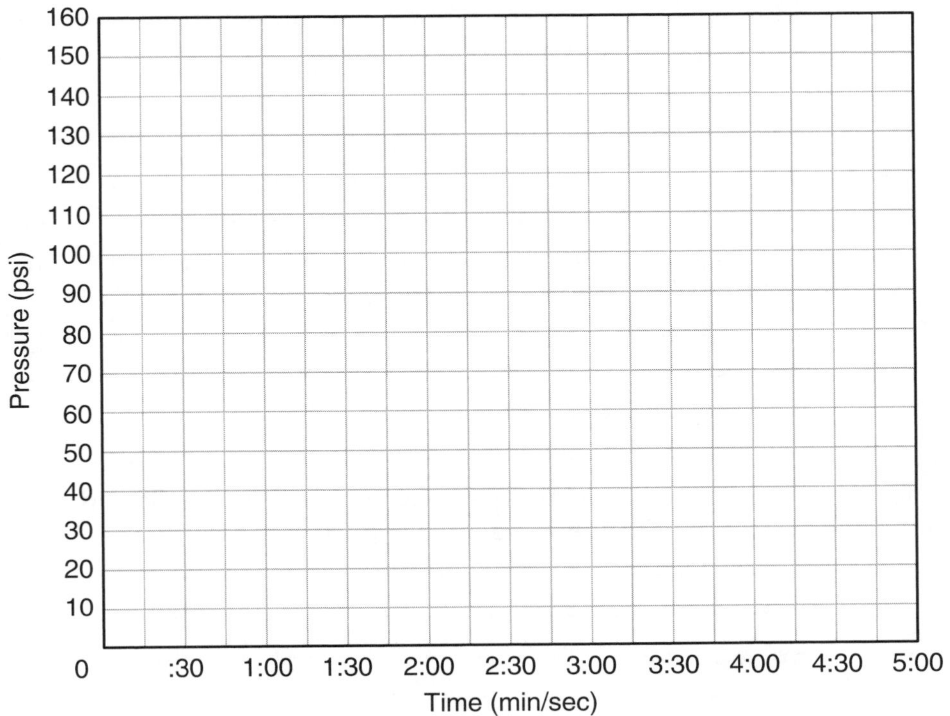

2. What is the maximum operating pressure difference between the receiver section of the system and the test circuit? What is the minimum operating pressure difference?

3. Does the range of the pressure differences identified in number 2 influence the performance of the test portion of the circuit? Justify your answer.

4. How does the shape of the plotted line representing air pressure in the receiver change as the airflow through the test circuit increases? Why?

5. How does the shape of the plotted line representing air pressure in the workstation line change as the airflow through the test circuit increases? Why?

____

____

____

____

6. How consistent is the compressor stop pressure and restart pressure throughout this activity? Use observed pressures to support your answers.

____

____

____

____

7. In an actual operating system, how consistent would the elapsed time be between compressor restart and compressor stop? What factors would influence this time?

____

____

____

____

8. What happened when the flow control valve was opened fully in steps 35 and 36? Explain the changes in system pressure and the operating cycle of the compressor.

____

____

____

____

9. In your judgment, can stop-start capacity control be considered effective for low-capacity compressors? Justify your answer by discussing information related to system performance, system cost, and system maintenance.

____

____

____

____

Activity 15-5

Estimate Required Compressor Capacity

Name ______________________________ **Date** ______________

Selecting a compressor may be relatively easy for small applications that have a limited number of workstations. However, large systems with a variety of tools that intermittently function require careful analysis. If the compressor is for a new installation, the process is further complicated by the need for an estimate of future system growth. This activity is designed to illustrate a method that may be used to estimate only the compressor capacity when selecting a unit for use in a new existing system.

Activity Specifications

Estimate the required capacity of a compressor for a pneumatic system applying the following suggested procedures and the number and sizes of actuators, system operating pressures, and machine operating cycles designated by your instructor. Use the forms and the process suggested in the procedures section of this activity to determine required compressor capacity. Consult compressor manufacturer data sheets and any additional recommended formulas to determine the type and size of a compressor unit suitable for the designated system. Carefully analyze the data and your compressor recommendation before answering the activity questions.

Procedure to Determine Required Compressor Capacity

Selecting the proper size of a compressor for a pneumatic system depends on a number of factors. These factors are constantly changing in most systems. The unique operating characteristics of the air-powered tools in each system complicates the process.

Factors to consider

The following factors must be considered to assure the selection and installation of a compressor that is not excessively large, can adequately supply current needs, and provide a reserve capacity for increased air demand expected in the future.

1. Tool air requirement. The volume of air required by a tool to operate at full-load conditions.
2. Tool load factor (TLF). The ratio of the actual air consumption of a tool to the volume expected under continuous, full-load conditions. Expressed as a percentage.

   $TLF = TF \times WF$

   where:

   Time factor (TF) = The portion of a work period during which a tool is actually operating. Expressed as a percentage.

   Work factor (WF) = The air consumption of a tool under actual work conditions compared to consumption under full-load conditions.
3. Receiver air pressure. The maximum air pressure in the receiver and distribution lines.
4. Workstation air pressure. The maximum working air pressure in the lines leading from the FRL at the workstation to the individual air tools in the pneumatic system. These pressures will be set at a number of different levels, depending on the type of tools and functions planned for each station.

Suggested steps for calculations

Recommending a compressor for a large installation must include calculations based on information available from the compressor manufacturer and input from experienced engineers. This activity provides a guide to selection that identifies only the basic factors that must be considered when determining compressor size. This is only the first step in a process to select a complete compressor package. The texts recommends a seven-step process to determine the best package for an installation.

1. Determine total air requirements for each of the tools at each of the workstations. Assume the tool is continuously operating under full-load conditions.
2. Arbitrarily assign a time factor and a work factor for each of the tools.
3. Calculate a tool load factor for *each* of the tool types at *each* of the workstations.
   TLF = TF × WF
4. Determine the total air consumed by each of the workstations at the workstation pressure setting.
5. Convert each of the workstation air-consumption rates to a consumption rate at receiver and distribution line pressure.
6. Total the converted consumption rates to determine the volume of compressed air at receiver and distribution-line pressure needed to theoretically maintain the system pressure.
7. Increase the calculated total air consumption rate by 20% to 40% to assure the compressor can supply adequate air through unanticipated high load levels. The selected increase depends on the type of compressor and the tasks performed by the system.
8. Identify any imminent load increases related to circuit modification, increased production rates, or system expansion. Add an appropriate amount of consumption to the consumption rate calculated in previous steps.
9. Select a compressor capable of producing the needed volume of compressed air at the required system pressure.

Activity Exercise Problem

This section of the activity involves analysis of the compressed-air needs of a planned pneumatic system in order to establish the size of the compressor. Your instructor will supply details about the system, including descriptions of the various branches of the air distribution system and a listing of the air-powered machines and tools. Follow the nine steps described above. Calculate the amount of air needed to operate the various sections of the system as described.

It will be necessary for you to make some judgment calls regarding the operating of the various tools and circuit functions. For example, the distribution line pressure, time factor and work factor for determining the tool load factor, and the amount of increase in the air consumption to compensate for unanticipated load levels must be estimated. Be certain you understand the concepts involved and can justify the levels you select.

Show all formulas and calculations used to determine the amount of air needed to operate the system. Organize the calculations around the grouping of tools and equipment shown in the following charts.

Selected distribution line pressure ____

Fabrication area

Tool description	Workstation pressure	Rated air requirement	Number of tools	Load factor	Time factor

Cleaning area

Tool description	Workstation pressure	Rated air requirement	Number of tools	Load factor	Time factor

Assembly

Tool description	Workstation pressure	Rated air requirement	Number of tools	Load factor	Time factor

Finishing and painting

Tool description	Workstation pressure	Rated air requirement	Number of tools	Load factor	Time factor

Activity Analysis

Complete the following questions in relation to the data collected and calculations required to recommend the size for the system compressor.

1. Justify the operating pressure selected for the distribution line before beginning the calculations.

____

____

____

____

2. How were the rated air requirements obtained for the various equipment and tools? Were the units of measure used by the different manufacturers uniform or was it necessary to convert data in order to make calculations?

3. How important is the load factor in these calculations? Why? What error would result in compressor sizing if the load factor is too low or too high?

4. Describe the procedure you used to establish the time and work factors for the calculations.

5. What percentage was added to the initially calculated, total air-consumption rate? Which factors influenced you in selecting this percentage?

6. Identify two factors that encourage and two factors that discourage installing a compressor that is larger than currently needed.

7. Show all formulas and calculations required to establish the recommended size of compressor for this installation.

Chapter 16

Conditioning and Distribution of Compressed Air

Controlling Dirt, Moisture, Temperature, and Pressure

Air for use in a pneumatic system is readily available from the atmosphere that covers Earth to a depth of several miles. However, this huge resource of air must be conditioned, compressed, distributed, and controlled to be effectively used in a pneumatic system. This process is often complicated by factors such as air contaminates, air storage and distribution restrictions, and the control demands of system applications. Each of these conditions must be considered with the application of appropriate components to assure effective system operation.

These activities are designed to illustrate how compressed air is stored and distributed in an operating pneumatic system and cover the sizing and operation of pressure control components located at the system workstations.

> **Note:** Check with your instructor to be certain that all the policies of the school are followed, especially if your activities involve a group other than the school.

Key Terms

The following terms are used in this chapter. As you read the text, record the meaning and importance of each. Additionally, you may use other sources, such as manufacturer literature, an encyclopedia, or the Internet, to obtain more information.

aftercooler ______________________________

____

____

air-distribution system ______________________________

____

____

air-drying equipment ______________________________

____

____

auxiliary air receiver ______________________________

____

____

balanced–poppet valve regulator______________________________

____

____

centralized grid ______________________________

____

____

decentralized grid ______________________________

____

____

desiccant ______________________________

____

____

dew point______________________________

____

____

diaphragm-chamber regulator______________________________

____

____

direct-operated regulator ______________________________

____

____

draincock______________________________

____

____

drop lines ______________________________
____
____

dry filters ______________________________
____
____

filter ______________________________
____
____

flexible conductor ______________________________
____
____

flexible hoses ______________________________
____
____

FRL unit ______________________________
____
____

humidity ______________________________
____
____

intake-line filter ______________________________
____
____

intercooler ______________________________
____
____

load factor ______________________________
____
____

loop system ______________________________
____
____

lubricator ______________________________
____
____

moisture separator ______________________________
____
____

oil-bath filters ______________________________
____
____

oil-wetted filters ______________________________
____
____

pilot-operated regulator ______________________________
____
____

pressure regulator ______________________________
____
____

relieving-type regulator ______________________________
____
____

rigid conductor ______________________________
____
____

saturation point ______________________________
____
____

time factor ______________________________
____
____

water trap ______________________________
____
____

Chapter 16 Quiz

Name ______________________ Date ____________ Score ________

Write the best answer to each of the following questions in the blanks provided.

1. List three considerations for the system air intake.

____

____

____

2. Compressing air produces ______, which increases both the temperature and pressure of the air in a pneumatic system.

____

3. What is the purpose of an intercooler? What is the purpose of an aftercooler?

____

____

____

____

4. What determines how much water vapor air can hold?

____

____

____

____

5. Name three methods used by air drying equipment to remove water vapor from compressed air.

____

____

____

6. Some manufacturers provide a formula for sizing a pneumatic system receiver, but suggest that the result be multiplied by a factor of ______ to ______ to determine actual capacity.

____

7. ______ grid air-distribution systems use two or more independently operating compressors and grids.

____

8. The initial sizing and layout of a compressed-air-distribution system should consider an increase in air demand and the ______ of system actuators.

____

9. Time and motion studies can produce an actuator operating ______ for use in establishing estimated air consumption.

____

10. Air-distribution lines should be designed with a minimum slope of one inch per ______ feet of line to allow water to accumulate at selected low points in the system.

____

11. Name the three functions provided by the final air preparation unit located at the workstation of a pneumatic system.

____

____

____

____

____

____

12. Water and dirt that accumulate at the bottom of the filter bowl are removed by an automatically operating device or a manually opened valve called a(n) ______.

____

13. The type of pressure regulator selected for a particular situation depends on the ______ and control needed for an application.

____

14. What is the purpose of the transfer pin in direct-operated and basic, diaphragm-chamber regulators and where is it located?

____

____

____

____

15. A relieving-type pressure regulator uses a(n) ______ to prevent external forces from increasing the regulator outlet pressure above the setting of the unit.

____

16. The pressure in the pilot-air chamber of a pilot-operated pressure regulator is controlled by a small, ______ located at a remote location.

____

17. Oil is added to workstation compressed air to improve the ______ and ______ of pneumatic system valves, actuators, and tools.

____

18. To prevent excessive pressure drop during periods of high air use, lubricators usually have a(n) ______ valve that allows system air to directly enter the working lines.

____

19. Pressure loss figures for air passing through conductors is commonly expressed in loss per ______ feet of a specific conductor.

____

20. At a minimum, hose is comprised of three layers. What is the purpose of each layer?

____

____

____

Activity 16-1

Analyze the Construction and Operation of a Compressed-Air Storage and Distribution System

Name ______________________________ **Date** ______________

This activity is designed to show the structure of air storage and distribution in an operating pneumatic system. The examination will begin with the receiver and end with the workstation. The activity requires comparison of suggested, ideal distribution line layout and component applications with an actual working system.

Activity Specifications

This activity involves the examination and analysis of an operating air storage and distribution system. Visit a facility to observe the layout and construction of the distribution system and lines, as well as the overall operation of the equipment powered by the system.

Site Assignment and Preparation for Site Visit

Consult your instructor to identify possible locations for this activity. Be certain to obtain permission to visit the facility. When you request permission, indicate you will follow all safety regulations required by the organization that owns and operates the system.

Use the questions in the activity analysis portion of this activity to help structure the visit to the assigned system. Review the questions before the visit to be certain you obtain all of the information needed.

Activity Analysis

Complete the following questions in relation to the observed distribution system.

1. Describe the physical facility in which the distribution system is located. Is the distribution system a part of the original building construction or was it added as the work environment was modified?

   ____
   ____
   ____
   ____

2. Describe the type of work being done by the pneumatic equipment in the facility. Does the distribution system appear to adequately meet the air needs of the tools and equipment performing the work? Support your answer with examples based on your observations.

   ____
   ____
   ____
   ____

3. Estimate the number of individual workstations in the facility. Estimate the percentage of machine functions operated by the pneumatic system. How important are these functions to the overall operation of the facility?

4. Describe the size, shape, and construction of the receiver used in the distribution system. Determine the operating range of air pressure in the receiver. How is this operating range set and controlled for the system?

5. Develop a pictorial diagram showing the layout of the air-distribution system for the facility. Compare your diagram with the recommended styles suggested by manufacturers of pneumatic components. Discuss the variations you find.

6. Identify the methods used in the system to allow the removal of water from the lines. Describe the location of each in the system.

7. Describe any examples of air leakage observed in the system. How could the number of leaks be reduced? Would the savings from reducing this leakage justify the costs involved? Explain.

8. What types of conductors are used to connect the various segments of the system? What types of flexible connectors are used between pneumatic components and moveable machine parts?

9. Discuss the effectiveness of the air distribution in this system. What changes could be made to improve the system? Would the changes be cost effective? Explain.

Activity 16-2

Test and Compare the Performance of Direct-Operated and Relieving-Type Pneumatic Pressure Regulators

Name ______________________________ **Date** ____________

The pressure of the air used at the workstation is controlled by a pressure regulator. Regulators reduce high pressure in the distribution line to a lower pressure for operating workstation circuits and tools. The regulator is typically part of the final air preparation units used in pneumatic systems. Several different designs and features are available for these regulators. This activity involves operating and testing pressure regulators to illustrate the ability of the components to maintain a selected operating pressure.

Activity Specifications

Construct and operate a pneumatic circuit to test the operation of direct-operated and relieving-type pressure regulators. Use the test circuit and procedures shown below to set up and operate the circuit. Collect the data specified in the Data and Observation Records section of the activity. Analyze the circuit and collected data before answering the activity questions.

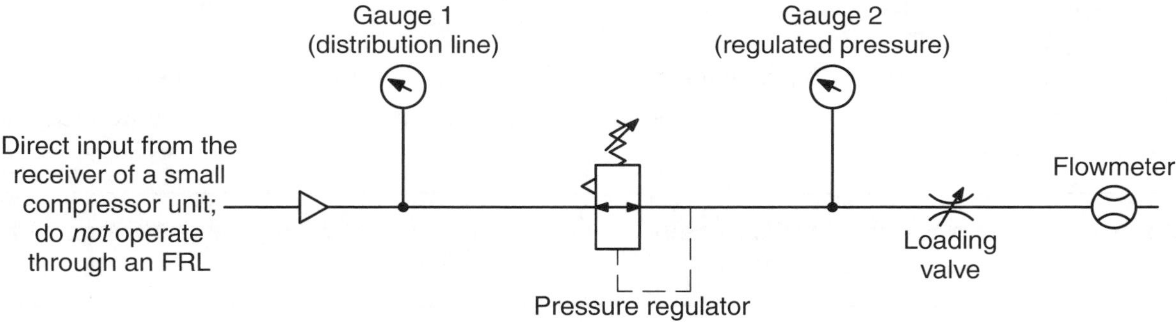

Required Components

Access to the following components is needed to complete this activity.

- Small, pneumatic compressor unit to simulate a large compressor station and air distribution system (must include compressor, receiver, safety valve, and pressure switch).
- Direct-operated pneumatic regulator (used to test the performance of the regulator design).

- Relieving-type pneumatic regulator (used to test the performance of the regulator design).
- Adjustable flow control valve (used to simulate a workstation load).
- Two pressure gauges (one used to measure distribution line pressure and the second to measure the regulated pressure).
- Flowmeter (used to measure the flow rate of air exhausting to the atmosphere).
- Manifolds/connectors (adequate number to make component connections).
- Hoses (adequate number to allow assembly of the test system).

Procedures

Study the following procedures to become familiar with the steps needed to complete the activity. Then, complete the procedures and record the observations indicated in the Data and Observation Records section.

1. Select the components needed to assemble the test circuit.
2. Assemble the system using the direct-operated regulator. Have your instructor check your selection of components and the setup of the circuit before proceeding.
3. Operate the compressor unit to determine the stop and restart pressure settings. Record these in the Data and Observation Records section.

Establishing initial air pressure for the test circuit

4. Begin the test with the compressor unit stopped.
5. Adjust the pressure regulator so no air pressure will be applied to the regulated side (gauge 2) of the pressure regulator.
6. Close the flow control valve to prevent air from being exhausted to the atmosphere.
7. Start the compressor and allow receiver pressure to increase until the pressure switch functions to stop the prime mover.

> **Note:** Record the highest and lowest distribution-line pressures displayed on gauge 1 for each of the loading valve settings made during the following performance tests.

Performance testing of the direct-operated regulator

8. Adjust the pressure regulator until gauge 2 reads 60 psi.
9. Record in chart 1 in the Data and Observation Records section the pressure on gauges 1 and 2 and the airflow through the flowmeter.
10. Open the flow control valve 1/4 turn from the seated position. Record the pressure on gauges 1 and 2 and the airflow through the flowmeter.
11. Repeat step 10 for each of the flow control valve settings shown in chart 1.

> **Note:** Turn the loading valve slowly when completely opening the flow control valve.

12. Adjust the pressure regulator so no air pressure is applied to the regulated side (gauge 2) of the pressure regulator.
13. Close the flow control valve to prevent air from being exhausted to the atmosphere.
14. Readjust the pressure regulator until gauge 2 reads 100 psi.
15. Record in chart 2 in the Data and Observation Records section the pressure on gauges 1 and 2 and the airflow through the flowmeter.

16. Open the flow control valve 1/4 turn from the seated position. Record the pressure on gauges 1 and 2 and the airflow through the flowmeter.
17. Repeat step 16 for each of the flow control valve settings shown in chart 2.
18. Turn off power to the prime mover of the compressor.
19. Open the flow control valve and allow the system to exhaust all air to the atmosphere.

Performance testing of the relieving-type regulator

20. Replace the direct-operated pressure regulator with the relieving-type regulator.
21. Repeat steps 4 through 19 to test the performance of the relieving-type pressure regulator. Record data and general observations in charts 3 and 4 in the Data and Observation Records section.

Activity completion and component cleanup and storage

22. Discuss your data and observations with your instructor before disassembling the test circuit.
23. Clean and return all components to their assigned storage locations.
24. Complete the activity questions.

Data and Observation Records

Record pressure readings, airflow rates, and general observations concerning the operation of the direct-operated and relieving-type pressure regulators.

Compressor stop pressure: ____________________

Compressor restart pressure: ____________________

Chart 1 Direct-Operated Pressure Regulator (60 psi setting)

Flow control/loading valve (turns open)	Line pressure (gauge 1)		Regulated pressure (gauge 2)	Airflow (cfm)
	Lowest	Highest		
Fully closed				
1/4				
1/2				
3/4				
1				
1 1/4				
1 1/2				
1 3/4				
2				
2 1/4				
2 1/2				
2 3/4				
3				
Fully open				

Chart 2 Direct-Operated Pressure Regulator (100 psi setting)				
Flow control/loading valve (turns open)	Line pressure (gauge 1)		Regulated pressure (gauge 2)	Airflow (cfm)
	Lowest	Highest		
Fully closed				
1/4				
1/2				
3/4				
1				
1 1/4				
1 1/2				
1 3/4				
2				
2 1/4				
2 1/2				
2 3/4				
3				
Fully open				

Chart 3 Relieving-Type Pressure Regulator (60 psi setting)				
Flow control/loading valve (turns open)	Line pressure (gauge 1)		Regulated pressure (gauge 2)	Airflow (cfm)
	Lowest	Highest		
Fully closed				
1/4				
1/2				
3/4				
1				
1 1/4				
1 1/2				
1 3/4				
2				
2 1/4				
2 1/2				
2 3/4				
3				
Fully open				

Chart 4 Relieving-Type Pressure Regulator (100 psi setting)				
Flow control/loading valve (turns open)	Line pressure (gauge 1)		Regulated pressure (gauge 2)	Airflow (cfm)
	Lowest	Highest		
Fully closed				
1/4				
1/2				
3/4				
1				
1 1/4				
1 1/2				
1 3/4				
2				
2 1/4				
2 1/2				
2 3/4				
3				
Fully open				

Activity Analysis

Complete the following questions and activities in relation to the data collected and observations of the test procedures.

1. Generally, how well do the regulators maintain a constant pressure level in the regulated (working) line between the regulator and load?

   ____

   ____

   ____

   ____

2. Compare the performance of the direct-operated and the relieving-type regulators. Identify at least one advantage and one disadvantage of each valve design.

   ____

   ____

   ____

   ____

3. What happens to the performance of a pneumatic actuator if the pressure regulator is set at a pressure higher than the restart pressure of the compressor?

   ____

   ____

   ____

   ____

4. Explain the pressure variations on gauge 1, as shown in charts 1 through 4, collected during the four performance tests.

5. What would happen to the performance of a regulator if the load's airflow demand exceeds the output of the compressor (distribution system)?

Chapter 17

Work Performers of Pneumatic Systems

Cylinders, Motors, and Other Devices

Pneumatic actuators include a wide variety of cylinders, motors, and power tools. Each of these actuators is available in a variety of designs and sizes. Actuators vary from precision, miniature cylinders designed to operate control circuits to large cylinders, motors, and impact tools used in manufacturing, construction, and mining operations. In recent years, use has grown to include applications in the general consumer market. Many actuators are off-the-shelf components incorporated in a variety of systems, while others are an integral part of a tool.

These activities are designed to illustrate the typical information available on data sheets supplied by actuator and tool manufacturers. Also included is the application of an approach for estimating the quantity of air consumed by actuators in an operating system.

Key Terms

The following terms are used in this chapter. As you read the text, record the meaning and importance of each. Additionally, you may use other sources, such as manufacturer literature, an encyclopedia, or the Internet, to obtain more information.

air consumption ____________________

air nozzles ____________________

blowguns ____________________

chipping hammers ____________________

corrosion resistance ____________________

double-acting cylinders ____________________

drying ____________________

force ____________________

gripper ____________________

impact wrenches ____________________

linear actuators ____________________

material agitation ____________________

material transfer ____________________

metal rolling ____________________

nail drivers ____________________

nutsetters ____________________

paving breaker ______________________________

____

____

piston motors ______________________________

____

____

rammers______________________________

____

____

reciprocating pneumatic motors ______________

____

____

resilient materials___________________________

____

____

riveting hammers____________________________

____

____

rock drills ______________________________

____

____

rotary actuators ____________________________

____

____

scaling hammers ____________________________

____

____

single-acting cylinders_______________________

____

____

spraying______________________________

____

____

tampers ______________________________

____

____

turbine motors ______________________________

____

____

vane motors______________________________

____

____

Chapter 17 Quiz

Name ______________________________ Date ____________ Score ________

Write the best answer to each of the following questions in the blanks provided.

1. What is found in compressed air that produces corrosive conditions requiring the use of chrome and other special materials in component construction?

____

2. Pneumatic cylinders that produce force during both extension and retraction are called ______ cylinders.

____

3. Name the two factors that must be considered to be certain a pneumatic cylinder can perform as expected in an operating circuit.

____

____

4. Name two factors that can cause inaccuracies when calculating the air-consumption rate of pneumatic cylinders.

____

____

5. What controls the maximum torque output of an air motor?

____

____

____

____

6. Which type of air motor is considered the most common?

____

7. Vane air motors are available with operating speeds as low as 100 rpm to over ______ rpm.

____

8. Piston air motors are available in both axial- and ______-piston designs.

____

9. How do turbine air motors produce a high air velocity to turn the turbine?

____

____

____

____

10. Selecting an air motor for a given situation requires careful analysis of the ______ that will be encountered.

____

11. Name the two techniques used for energy transfer in reciprocating air motors.

____

____

12. How is compressed air used in process functions such as spraying, drying, and material agitation?

____

____

____

____

13. List two safety factors related to blowguns.

____

____

14. In many designs of air nailers, how is the piston retracted to restart the cycle?

____

____

____

____

15. Pneumatic ______ are commonly used to bore holes for the placement of explosives in road-construction, quarrying, and mining operations.

____

Activity 17-1

Analysis of Pneumatic Cylinder Information Sheets

Name ______________________________ **Date** ______________

This activity is designed to show how manufacturers of pneumatic cylinders present technical information about the cylinders they manufacture. Most companies supply considerable information about the cylinders they produce. However, the exact content and presentation vary.

Activity Specifications

Analyze the data sheets of two cylinders from different manufacturers as designated by your instructor. The first model should have threads or rolled metal to attach the ends to the cylinder tube. The second model should use tie-rod construction. Carefully study the data sheets and then answer the activity questions. The information available for cylinders often includes considerable data on a large number of options.

Analyzed Pneumatic Cylinders

The cylinders will be designated by your instructor. Record the manufacturer and model of each cylinder.

Threaded or rolled-end construction

Manufacturer ______________________________

Model ______________________________

Tie-rod construction

Manufacturer ______________________________

Model ______________________________

General Features and Specifications of Cylinder Models

Obtain the following information from the website or printed catalog of each cylinder manufacturer. The information includes basic factors commonly found in the literature. However, some companies include many more details than shown here.

Threaded or rolled-end construction cylinder

Bore range: ______________________________

Stroke range: ______________________________

Available rod diameters: ______________________________

Maximum operating pressure: ______________________________

Operating temperature range: ______________________________

Available port sizes: ______________________________

Tie-rod construction cylinder

Bore range: ______

Stroke range: ______

Available rod diameters: ______

Maximum operating pressure: ______

Operating temperature range: ______

Available port sizes: ______

Activity Analysis

In this section, you must analyze the information gathered in the previous section. Further study of the catalogs, service manuals, and the websites of the companies may also be needed. Pay close attention to service and maintenance information included in the materials. Some of the following questions require you to be aware of these issues, as well as the specifications and construction details of the cylinders.

1. Compare the bore and stroke ranges of the threaded or rolled-end cylinder and the tie-rod cylinder. Which type appears to provide the larger selection of sizes? Why do you think this is true?

______

2. Compare the pressure ratings of cylinders. Are the ratings similar or different? Why?

______

3. Discuss the construction differences you can easily identify in the cylinder models.

______

4. Identify the features in each cylinder that dampen the noise produced as the piston reaches the end of the stroke. Describe any variations in cushion design between the two assigned cylinders.

______

5. How many different mounting styles are available for the threaded or rolled-end cylinder? List them.

______

6. How many different mounting styles are available for the tie-rod cylinder? List them.

______

7. Examine the seal designs used in the construction of these cylinders. Where are the seals located and what materials are used for their composition?

8. What type of lubrication is needed by each of these cylinders?

9. What type of cylinder accessories (mounting plates, rod clevis, etc.) are available for each of these cylinders?

10. What happens to system performance and service when side-loading of the cylinder is *not* considered? What can be done during cylinder installation to reduce or eliminate side loading?

Activity 17-2

Estimate Cylinder Air Consumption in an Operating Pneumatic Circuit

Name ______________________________ **Date** ______________

The volume of air consumed by actuators is directly related to the effective operation of a pneumatic system. A system will not properly function if the air needed for an actuator cycling speed approaches the capacity of the compressor or distribution system. The many variables involved in an operating pneumatic system makes it difficult to determine the required volume of air. This activity applies a method that may be used to estimate air consumption of pneumatic cylinders in an operating system.

Activity Specifications

Estimate the volume of air needed to operate the cylinders in a pneumatic circuit designated by your instructor. Use the formula suggested in the text and shown in the next section. Take into consideration factors such as air leakage in the system. Also, consider the use of a correction factor to assure an adequate supply of air. Carefully review your calculations and justify any adjustments for temperature variations and leakage or the use of a correction factor before answering the activity questions.

Determining Air Consumption

Effective operation of a circuit requires adequate air to maintain the pressure and flow rate needed to perform required tasks. When adding actuators to a system, it is critical to calculate the additional load being placed on the compressor and distribution system. Be certain the new air requirement added to the existing load demand does not exceed system capacity. Overloading either the compressor or distribution lines will reduce system efficiency.

The air consumption of a cylinder in an operating circuit can be estimated using the formula:

$$CFM = V \times Pr \times N$$

where:

CFM = Cubic feet of free air required per minute of operation

V = Volume of compressed air per cycle

$$= \frac{(Ab \times S) + (Ar \times S)}{1728}$$

where:

Ab = Cross-sectional area in square inches of the blind end of the cylinder piston

Ar = Effective cross-sectional area in square inches of the rod end of the cylinder piston

S = Length of cylinder stroke in inches

1728 = Cubic inches per cubic foot

Pr = Absolute pressure ratio

$$= \frac{\text{Operating psig} + 14.7}{14.7}$$

N = Number of cylinder cycles per minute

The above formula provides only an estimate of the air needed to operate a cylinder. Three factors that are difficult to control can cause inaccuracies in the results:

- Varying temperatures in an operating system.
- Air leaks in the circuit.
- Variations in the volume of the cylinder air chambers due to design factors such as cushions.

However, the calculation produces the best-available estimate. Experienced circuit designers often add a correction factor to the formula to compensate for various cylinder features, circuit designs, and load conditions.

Suggested steps of procedure for calculation

The process to determine cylinder air consumption is relatively simple. The following steps provide a general guide for calculations. When a number of different-sized cylinders are used in a machine, the air consumption for each cylinder needs to be separately calculated and then combined to identify total air demand of the machine.

1. Identify the cylinder model number.
2. Locate the cylinder in the manufacturer's catalog or on the company website.
3. Establish the cylinder bore, rod diameter, and stroke using manufacturer data or by physically measuring the cylinder.
4. Calculate the cubic inches of air needed to fully extend the cylinder ($Ab \times S$).
5. Calculate the cubic inches of air needed to fully retract the cylinder ($Ar \times S$).
6. Calculate the total amount of air needed per cylinder cycle ($Ab \times S$) + ($Ar \times S$).
7. Establish a correction factor to compensate for temperature variations, leaks, and variations in cylinder volume using information about the age of the pneumatic system, level of machine maintenance, and ambient temperatures.
8. When a machine contains multiple cylinders, calculate the air needed by each individual cylinder to obtain the total air supply needed to properly operate the machine.

Activity problem

This section of the activity involves analysis of the compressed-air needs of a machine being added to the production line of a manufacturing plant. The machine includes two clamping cylinders and two reciprocating cylinders that move finishing tools to deburr the edges of stamped-metal parts. Refer to the circuit schematic shown on the next page. The deburring process requires two passes of the finishing tools to provide the desired finish. The machine operates automatically. The operator only needs to stack parts on an intake table and remove hoppers of finished parts.

The following list includes the cylinder specifications and operating information needed to calculate both cylinder and total machine air consumption. Your instructor will supply the details about cylinder sizes, operating pressures, operating speeds, and cycle times.

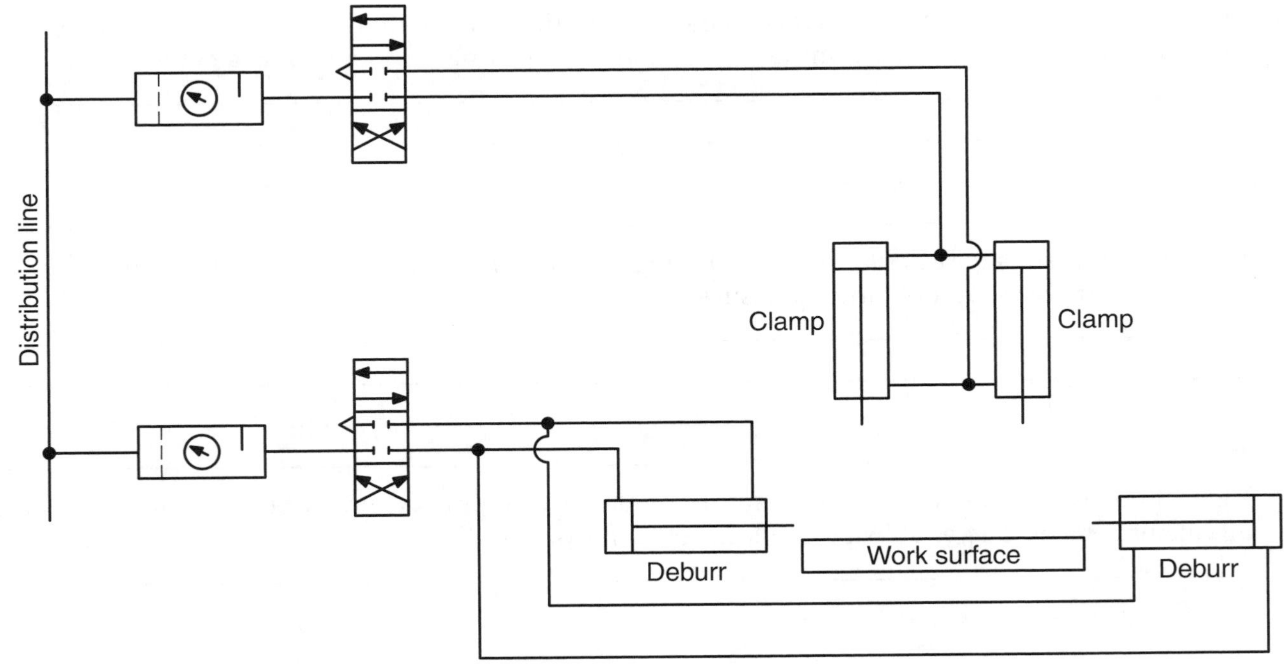

Clamping cylinders

Bore: ____________________

Rod diameter: ____________________

Stroke: ____________________

Operating pressure: ____________________

Cycle time: ____________________

Reciprocating cylinders

Bore: ____________________

Rod diameter: ____________________

Stroke: ____________________

Operating pressure: ____________________

Cycle time: ____________________

Calculate the amount of air needed to operate the various sections of the system as described. It will be necessary for you to make judgment calls regarding some of the operating aspects of the machine. Be certain you understand the concepts involved and can justify your decisions. Explain your approach, show all formulas and calculations, and answer the questions in the next section.

Activity Analysis

Complete the calculations and answer the questions in relation to the procedures recommended to estimate the volume of compressed air needed to operate the machine.

1. Define the term *absolute pressure ratio* (Pr) included in these calculations. Why must this factor be included in the formula to determine the volume of free air needed to operate a cylinder?

2. Explain the reasons for calculating the volume of needed free air, rather than determining the amount of air at system operating pressure.

3. Determine the amount of free air needed to operate the clamping cylinders. Show the formulas and all calculations used to establish the volume of air needed for this portion of the machine operation.

4. Determine the amount of free air needed to operate the reciprocating cylinders. Show the formulas and all calculations used to establish the volume of air needed for this portion of the machine operation.

____

5. Make a final recommendation concerning the volume of free air needed to operate the cylinders. Consider the need for a correction factor to adjust for temperature and air leaks. Include this correction factor in the recommended volume of air.

____

6. What correction factor did you add to the initial, calculated total air consumption rate? Which factors influenced you in selecting this percentage, rather than a more conservative or liberal figure?

____

____

____

____

Activity 17-3

Analysis of Pneumatic Motor Information Sheets

Name ______________________________ **Date** ______________

This activity is designed to show the data presented in information sheets provided by manufacturers of pneumatic motors. Most companies supply information about the construction, general operation, and performance of the motors they sell. The extent of information and the approach may vary, depending on the target market.

Activity Specifications

Analyze the data sheets of two pneumatic motors designated by your instructor. One of these motors should be designed to directly power a hand tool. The second motor should turn a gear-reduction unit that provides low-speed, high-torque output. The motors may have different capacities and be designed for different applications. Carefully study the information sheets and then answer the activity questions.

Analyzed Pneumatic Motors

The motors will be designated by your instructor. Record the manufacturer and model of each motor.

Direct-drive unit

Manufacturer: ______________________________

Model Number: ______________________________

Gear-drive unit

Manufacturer: ______________________________

Model Number: ______________________________

General Features and Specifications of Motor Units

Obtain the following information from the website or printed catalog of each motor manufacturer. The information includes basic factors commonly found in the literature. However, some companies include many more details than shown here.

Direct-drive motor

Size [length (in) × width (in) × height (in)]: ______________________________

Weight (lb): ______________________________

Maximum power (hp): ______________________________

No-load speed (rpm): ______________________________

Speed at maximum power (rpm): ______________________________

Torque at maximum power (lb/ft): ______________________________

Air consumption at maximum power (cfm): ______________________________

Line pressure used for rating motor (psi): ______________________________

Gear-drive motor unit

Size [length (in) × width (in) × height (in)]: ___

Weight (lb): ___

Maximum power (hp): ___

No-load speed (rpm): ___

Speed at maximum power (rpm): ___

Torque at maximum power (lb/ft): ___

Air consumption at maximum power (cfm): ___

Line pressure used for ratings unit (psi): ___

Activity Questions

In this section, you must analyze the information gathered in the previous section. Further study of the catalogs, service manuals, and the websites of the companies may also be needed. Pay close attention to service and maintenance information included in the materials. Some of the following questions require you to be aware of these issues, as well as the specifications and construction details of the motors.

1. List at least four characteristics that make pneumatic motors desirable for hand tools.

2. What motor design and pneumatic system factors produce high-speed rotation under no-load operating conditions?

3. What basic designs are used for the construction of the two pneumatic motors? What other motor designs are available for use with other applications?

4. Using resources from manufacturers, develop a definition for the term *starting torque* as it relates to pneumatic motors. Does this factor change with motors using other basic designs? Why or why not?

5. Describe how the performance of each motor will change as air pressure is increased at the workstation. Justify your answer.

6. Identify two specific reasons gears trains are used in conjunction with pneumatic motors.

____

____

7. Using resources from manufacturers, develop a definition for the term *throttling* as it relates to the design and operation of pneumatic motors used in hand tools.

____

____

____

____

8. Describe how the information included in the General Features and Specifications of Motor Units section of this activity can be used to select a pneumatic motor for a specific application. Name at least three other factors you feel should be considered during the selection process.

____

____

____

____

____

____

Activity 17-4

Analysis of Information Related to Pneumatic Tools

Name ______________________________ **Date** ____________

This activity is designed to familiarize you with the type of information available in catalogs, information sheets, and data specifications from manufacturers and distributors of pneumatic tools. These materials often provide information promoting the features of the equipment, as well as basic technical information about tool performance. These materials can be used to quickly compare the features and performance of a variety of tools.

Activity Specifications

Analyze the materials for three different types of pneumatic hand tools designated by your instructor. These tools should include a nailer, impact wrench, and grinder. The rating of the tools should be representative of the equipment typically found in construction and manufacturing applications. Carefully study the materials and answer the activity questions.

Analyzed Pneumatic Tools

The tools will be designated by your instructor. Record the manufacturer and model of each tool.

Nailer

Manufacturer: ______________________________

Model: ______________________________

Impact wrench

Manufacturer: ______________________________

Model: ______________________________

Grinder

Manufacturer: ______________________________

Model: ______________________________

General Tool Specifications and Operating Features

Obtain the following information from the website, printed catalog, or data sheet of each tool manufacturer. The information includes basic factors commonly found in the literature. However, some companies include many more details than shown here.

Nailer

Unit height: ______________________________

Unit length: ______________________________

Unit weight: ______________________________

Recommended operating pressure (psi): ______________________________

Air consumption (cfm): ______________________________

Driving force (in-lbs): ______________________________

Fasteners: ___

Type: ___

Gauge: ___

Magazine capacity: ___

Impact wrench

Unit height: ___

Unit length: ___

Unit weight: ___

Square drive size: ___

Recommended operating pressure (psi): ___

Air consumption (cfm): ___

Working torque range (ft-lbs): ___

Impacts per minute: ___

Maximum no-load speed (rpm): ___

Grinder

Unit length: ___

Unit weight: ___

Threaded spindle size: ___

Wheel maximum diameter: ___

Wheel maximum thickness: ___

Recommended operating pressure (psi): ___

Air consumption (cfm): ___

Maximum no-load speed (rpm): ___

Sound level (dB): ___

Activity Questions

In this section, you must analyze the information gathered in the previous section. Further study of the catalogs, service manuals, and the websites of the companies may also be needed. Pay close attention to warranty information, safety issues, and service and maintenance information included in the materials. Some of the following questions require you to be aware of these issues, as well as the specifications and construction details of the tools.

1. What is the pressure range typically used with these tools? Identify any tool that recommends an exceptionally high or low operating pressure. What appears to be the reason for this recommendation?

2. What type of devices are used to feed the nails used in the nailer. What other designs are used to hold the supply of nails in these devices?

3. What determines the driving power of the nailer? Include at least two factors.

4. Describe several safety factors listed in the operating manual of the pneumatic nailer. Why is safety such a major factor for these tools?

5. Describe how the desired torque level is set on the impact wrench.

6. Explain *impacts per minute* as it relates to an impact wrench.

7. Why should only specially designed sockets be used with impact wrenches? Refer to manufacturer literature.

8. Explain the term *free speed* as it relates to impact wrenches and grinders. Refer to manufacturer literature. Why is free speed so high in most of these tools?

9. What safety factor is a major concern when using a portable pneumatic grinder? What can be done to reduce injuries when using these grinding devices?

10. Describe the type of service and repair provided by the manufacturers of the tools. What variations exist among the companies?

11. Many manufacturers provide a means to obtain repair parts for the user to perform service. Describe the best service process used by the manufacturers.

Chapter 18

Controlling a Pneumatic System

Pressure, Direction, and Flow

In order to make efficient use of the compressed air, a series of control components must be incorporated into the pneumatic system. These components typically consist of valves providing pressure, direction, and speed control of actuators to obtain the desired force and movement of machine members. Control valves are typically located at workstations and in the circuits controlling system actuators. A variety of valve and circuit configurations are available to provide the required control. The following activities provide information on a technique to estimate the required size of a valve. The operation and application of control valves in basic circuits are also presented.

Key Terms

The following terms are used in this chapter. As you read the text, record the meaning and importance of each. Additionally, you may use other sources, such as manufacturer literature, an encyclopedia, or the Internet, to obtain more information.

bypass control system ______________________

check valve ______________________

compressor-capacity control ______________________

diaphragm ______________________

five-port directional control valve ______________________

fixed orifice ______________________

four-way directional control valve ______________________

meter-in circuit ______________________

meter-out circuit ______________________

mufflers ______________________

needle valve ______________________

packed-bore design ______________________

packed-spool design ______________________

piston ______________________

poppet valve ______________________

pressure booster ______________________

quick-exhaust valve ______________________

safety valve ______________________

shut-off valve

shuttle valve

sliding plate

spool

three-way directional control valve

timing-volume reservoir

valve body

Chapter 18 Quiz

Name ______________________ Date ____________ Score ________

Write the best answer to each of the following questions in the blanks provided.

1. The body of a control valve is made from standard bar stock or a(n) ______.

2. Describe a *diaphragm.*

3. What are the three pneumatic system sections that must provide specific pressure control if all phases of the system are to function effectively?

4. The device used to maintain a desired pressure at a pneumatic workstation is called an air ______.

5. An air-to-air pressure booster can be used if a section of a workstation circuit requires an air pressure ______ than the operating pressure of the system distribution lines.

6. List the four classifications of pneumatic directional control valves.

7. What allows free flow of air in one direction, but blocks air movement if the flow direction is reversed?

8. What is the primary purpose of three-way directional control valves?

9. List the three methods used to obtain a positive seal between the spool and bore in pneumatic four-way directional control valves.

10. The ______ -center configuration of a three-position, four-way directional control valve blocks the pressure port of the valve and connects both actuator ports to the exhaust.

11. Reverse airflow around pneumatic flow control valves is easily provided by the use of a properly positioned ______ valve.

____

12. In a flow-control circuit using a five-port, four-way directional control valve, where is the preferred location for the flow control valve?

____

13. Rapidly exhausting air from a cylinder increases extension or retraction ______ and increases the operating efficiency of the system.

____

14. Name the two functions of the mufflers commonly used on pneumatic valves and actuators.

____

____

15. List four reasons it is important to properly size pneumatic valves.

____

____

____

____

Activity 18-1

Construct and Test a Pneumatic Circuit Providing Multiple Pressure Levels

Name ______________________________ **Date** ______________

This activity is designed to examine the structure of the pressure control section of an operating pneumatic circuit. Pressure regulators are used to obtain several pressure levels. The activity allows a comparison of circuit configurations and regulator types that produce the desired pressure variations.

Activity Specifications

Construct and operate a pneumatic circuit that provides three pressure levels for machine operation. The circuit should be constructed using the components listed and the diagram shown below. Collect data as specified in the activity procedures section. Carefully analyze the circuits and the data collected and then answer the activity question.

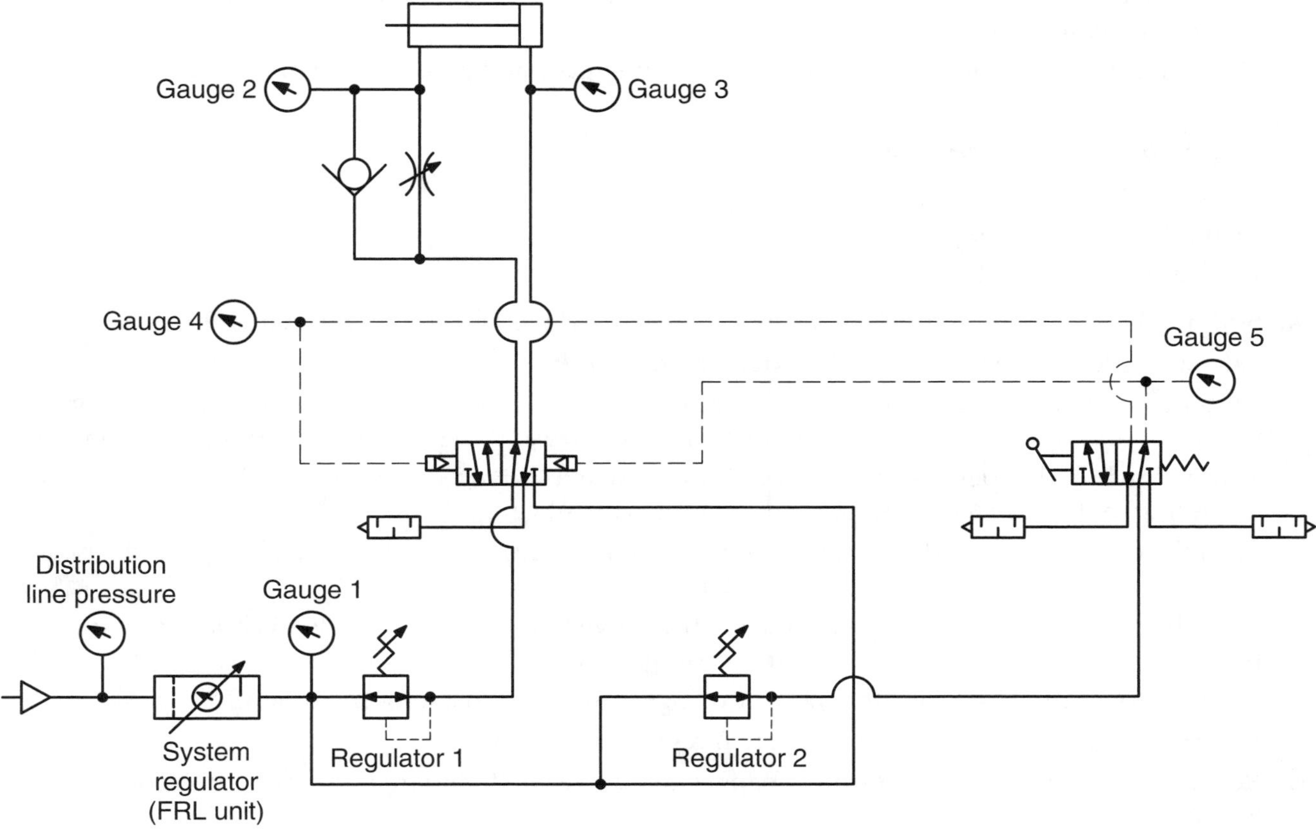

Required Components

Access to the following components as needed to complete this activity.

- Small pneumatic compressor unit or pneumatic distribution system workstation air supply.
- FRL unit (supplies high-pressure, regulated, conditioned air to the test circuit).

- Two direct-operated or relieving-type pneumatic regulators (used to provide additional pressure options to portions of the test circuit).
- Double-acting cylinder (used as the actuator to simulate circuit loading).
- Adjustable flow control valve with a built-in check valve (used to control the extension speed of the actuator).
- Pilot-operated, five-port, two-position, four-way directional control valve (used to control the direction of movement of the cylinder).
- Manually operated, five-port, two-position, four-way directional control valve with spring offset (used to shift the pilot-operated, four-way directional control valve).
- Six pressure gauges (used to measure working line and pilot line pressures).
- Manifolds/connectors (adequate number to make component connections).
- Hoses (adequate number to allow assembly of the test system).

Procedures

Study the following procedures to become familiar with the steps needed to complete the activity. Then, complete the procedures and record the observations indicated in the Data and Observation Records section.

1. Select the components needed to assemble the circuit.
2. Assemble the system. Have your instructor check your selection of components and the setup of the circuit before proceeding.
3. Start the compressor unit or turn on the main air supply and adjust the regulator valves to these pressures:

Regulator	**Pressure**
System regulator	60 psi
Regulator 1	30 psi
Regulator 2	45 psi

4. Shift the lever-operated directional control valve to extend and retract the cylinder.
5. Adjust the flow control valve so the extension time is approximately 5 seconds.
6. With the manually operated directional control in the normal position and the cylinder fully retracted, note and record the distribution line pressure and the pressure readings on all gauges.
7. Shift the manually operated directional control valve to extend the cylinder. While the cylinder is extending, note and record the pressure readings on all gauges.
8. With the cylinder fully extended, hold the directional control valve in the shifted position. Note and record the pressure readings on all gauges.
9. Shift the manually operated directional control valve to retract the cylinder. While the cylinder is retracting, note and record the pressure readings on all gauges.
10. Note any pressure variations that occur during cylinder extension or retraction.
11. Increase the pressure setting of the system regulator to 80 psi.
12. Repeat steps 4 through 10. Note any changes that were needed to maintain the extension time of 5 seconds.
13. Discuss your observations with your instructor.
14. Disassemble the circuit, wipe the components, and return them to their proper storage location.
15. Complete the activity questions.

Data and Observation Records

Distribution Line Pressure	Regulator Setting			Cylinder Movement	Gauge Pressure				
	1	2	System		#1	#2	#3	#4	#5
	30	45	60	Retracted					
				Extending					
				Extended					
				Retracting					
	30	45	80	Retracted					
				Extending					
				Extended					
				Retracting					

Activity Analysis

Answer the following questions based on the pressure readings and your observations during the operation of the circuit at the various pressure regulator settings.

1. Which circuit component controls the maximum pressure that can be obtained to retract the cylinder or shift the pilot-operated directional control valve? Why?

2. Explain the relationship between the gauge 2 and gauge 3 pressure readings as the cylinder is extended under a no-load condition. Why is this design of speed-control circuit preferred in pneumatic systems?

3. What advantage is gained in this circuit design by using a separate regulator to reduce the pilot pressure available to operate the directional control valve for the cylinder?

4. What advantage is gained in this circuit design by using a separate regulator to allow the cylinder retraction pressure to differ from the extension pressure?

5. Describe any pressure variations that occurred in the pilot and cylinder retraction portions of the circuit when the system regulator is increased from 60 psi to 80 psi in step 11.

6. Identify the source of any pressure indicated on gauge 3 during cylinder retraction.

7. Describe any variations in the pressure observed on the distribution line pressure gauge. What factors related to the distribution system could produce noticeable pressure variations? What system situations could result in pressure variations that go unnoticed?

Activity 18-2

Construct and Test Basic Pneumatic Speed-Control Circuits

Name ______________________________ Date ____________

Pneumatic actuator operating speeds are controlled by relatively simple flow control valves. These valves may be placed in different positions in a circuit to meter airflow into or out of the actuator. This activity is designed to show the configuration of basic pneumatic speed-control circuits and establish which valve location provides the most stable actuator movement.

Activity Specifications

Construct, operate, and analyze meter-in and meter-out pneumatic speed-control circuits. The activity is designed to illustrate the effectiveness of meter-in and meter-out flow control in pneumatic circuits. Use the test circuits and procedures shown below to set up and operate the circuits. Collect the data specified in the Data and Observation Records section of the activity. Analyze the circuits and collected data before answering the activity questions.

Meter-In Circuit

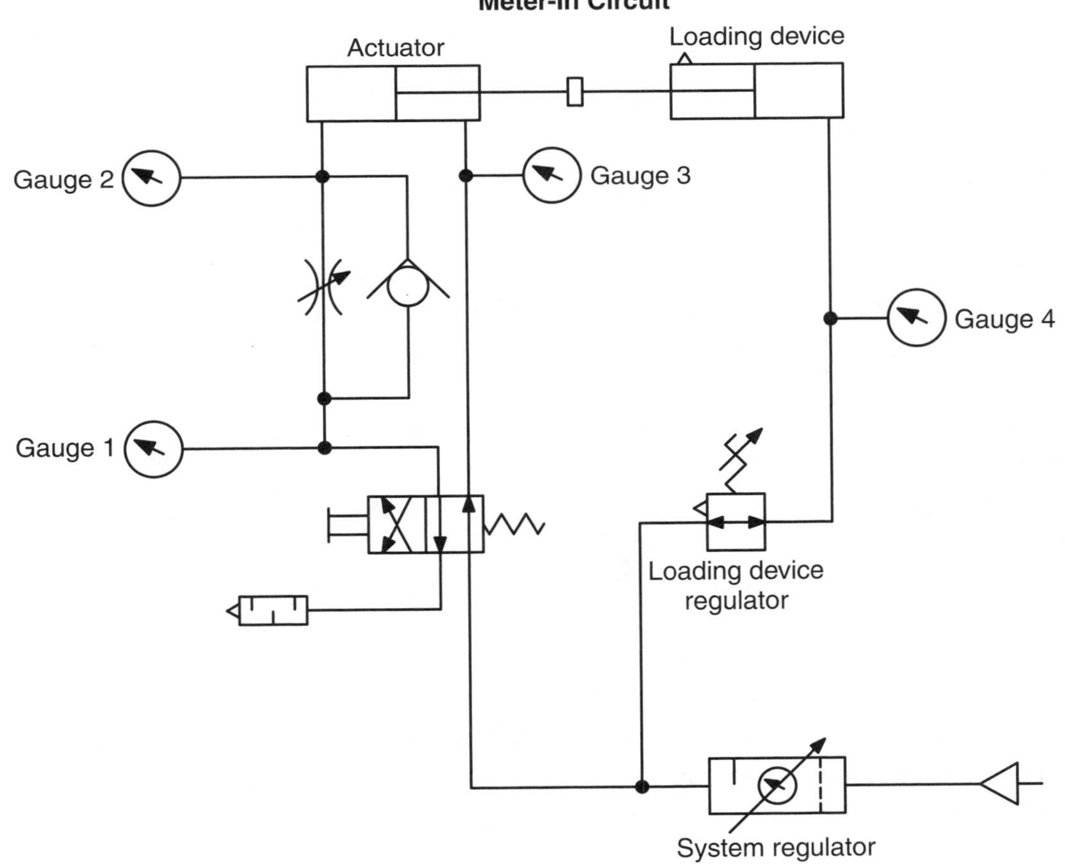

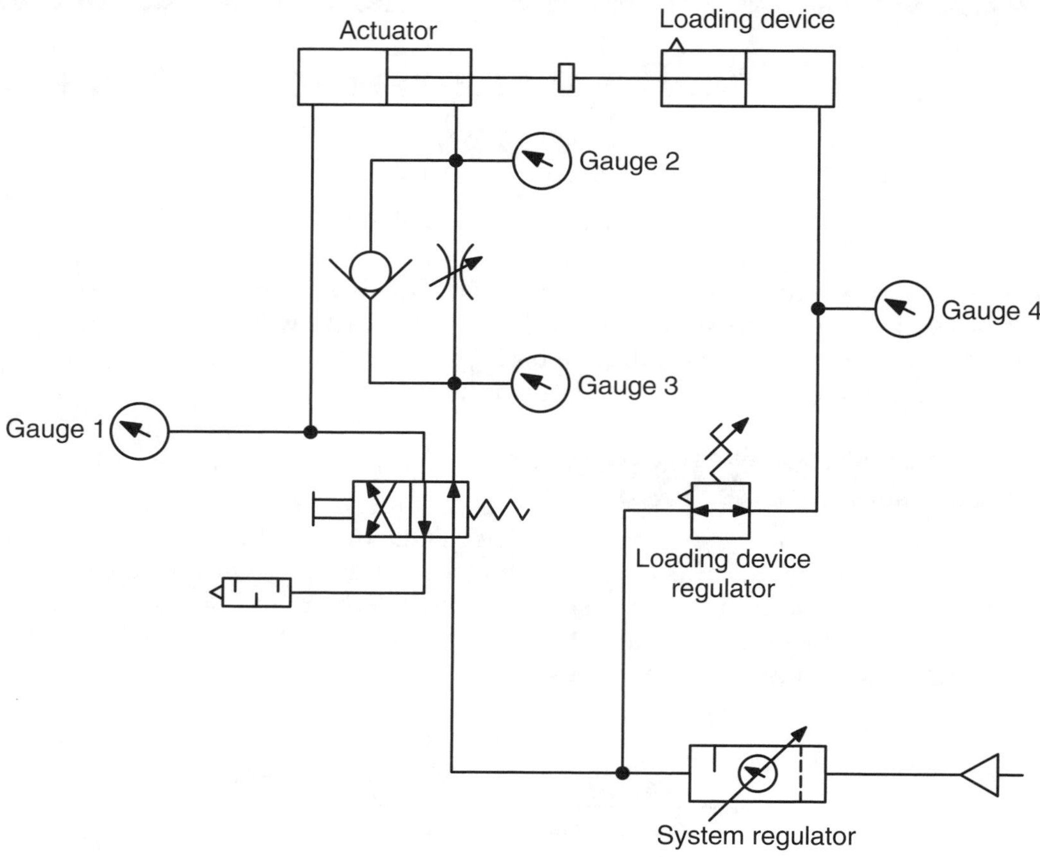

Required Components

Access to the following components is needed to complete this activity.

- Small pneumatic compressor unit or pneumatic distribution system workstation air supply.
- FRL unit (supplies high-pressure, regulated, conditioned air to the test circuit).
- Two double-acting cylinders (one used as the circuit actuator and the second used as a loading device).
- Mechanical coupling (used to connect the rod ends of the cylinders).
- Mounting platform (to hold cylinders in alignment).
- Relieving-type pneumatic regulator (used to control the pressure in the blind end of the loading cylinder).
- Adjustable flow control valve with a built-in check valve (used to control the extension speed of the circuit actuator).
- Manually operated, four-port, two-position, four-way directional control valve (used to control the direction of movement of the circuit actuator).
- Four pressure gauges (used to measure pressures in the working line and loading device).
- Manifolds/connectors (adequate number to make component connections).
- Hoses (adequate number to allow assembly of the test system).

Procedures

Study the following procedures to become familiar with the steps needed to complete the activity. Then, complete the procedures and record the observations indicated in the Data and Observation Records section.

1. Select the components needed to assemble the meter-in speed-control circuit.
2. Assemble the system.

> **Note:** Be certain the circuit actuator and loading cylinder are aligned and securely attached to the mounting platform. The rod ends of the cylinders must be securely connected using the mechanical coupling.

3. Have your instructor check your selection of components and the setup of the circuit before proceeding.

Meter-in speed control

4. Start the compressor unit or turn on the main air supply.
5. With the directional control valve in the normal position, adjust the FRL unit to 60 psi.
6. Adjust the loading device pressure regulator to 20 psi (gauge 4).
7. Shift the directional control valve to extend and retract the cylinder. Adjust the flow control valve so the cylinder extension time is approximately 5 seconds.
8. Extend the cylinder. Note and record in the chart in the Data and Observation Records section the cylinder extension time and the pressures registered on each of the four gauges.
9. Retract the cylinder. Note and record the cylinder retraction time and the pressures registered on each of the four gauges.
10. Extend the cylinder. While the cylinder is extending, vary the pressure setting of the loading device regulator. Observe any variations in actuator cylinder speed.
11. With the directional control valve in the normal position, adjust the loading device pressure regulator to 40 psi (gauge 4).
12. Repeat steps 8 through 10 for the 40 psi setting.
13. With the directional control valve in the normal position, adjust the loading device pressure regulator to 60 psi (gauge 4).
14. Repeat steps 8 thru 10 for the 60 psi setting.
15. Check the chart and the activity questions to be certain you have collected all of the information needed to answer the questions related to meter-in speed-control circuits. Repeat any sections of the activity as needed to clarify circuit operation.

Meter-out speed control

16. Turn off the air supply and modify the test circuit to create the meter-out speed-control circuit.
17. Start the compressor unit or turn on the main air supply.
18. With the directional control valve in the normal position, adjust the FRL unit to 60 psi.
19. Adjust the loading device pressure regulator to 20 psi (gauge 4).
20. Shift the directional control valve to extend and retract the cylinder. Adjust the flow control valve so that the cylinder extension time is approximately 5 seconds.
21. Extend the cylinder. Note and record the cylinder extension time and the pressures registered on each of the four gauges.

22. Retract the cylinder. Note and record the cylinder retraction time and the pressures registered on each of the four gauges.
23. Extend the cylinder. During cylinder extension, vary the pressure setting of the loading device regulator. Observe any variations in cylinder speed.
24. With the directional control valve in the normal position, adjust the loading device pressure regulator to 40 psi (gauge 4).
25. Repeat steps 21 thru 23 for the 40 psi setting.
26. With the directional control valve in the normal position, adjust the loading device pressure regulator to 60 psi (gauge 4).
27. Repeat steps 21 thru 23 for the 60 psi setting.
28. Check the chart and the activity questions to be certain you have collected all of the information needed to answer all questions. Repeat any sections of the activity as needed to clarify circuit operation.
29. Discuss your circuit observations with your instructor.
30. Disassemble the circuit, wipe the components, and return them to their proper storage location.
31. Complete the activity questions.

Data and Observation Records

Meter-in speed control

Loading Device Regulator Setting	Cylinder Movement		Gauge Pressure			
	Direction	Elapsed Time	#1	#2	#3	#4
20 psi	Extension					
	Retraction					
40 psi	Extension					
	Retraction					
60 psi	Extension					
	Retraction					

Meter-out speed control

Loading Device Regulator Setting	Cylinder Movement		Gauge Pressure			
	Direction	Elapsed Time	#1	#2	#3	#4
20 psi	Extension					
	Retraction					
40 psi	Extension					
	Retraction					
60 psi	Extension					
	Retraction					

Activity Analysis

Answer the following questions based on the pressure readings and your observations during the operation of the circuit at the various pressure regulator settings.

1. Why does the pressure in the blind end of the loading device cylinder (gauge 4) remain relatively constant during both the extension and retraction of the actuator cylinder? Where is the air exhausted as the cylinder piston in the loading device is forced to retract?

2. In the meter-in speed-control circuit, what causes the pressure on gauge 2 as the actuator cylinder is extended?

3. In the meter-out speed-control circuit, what causes the pressure on gauge 2 as the actuator cylinder is extended?

4. Explain any differences found between the gauge 1 pressures in the meter-in and meter-out circuits during extension of the actuator cylinder.

5. In the meter-in circuit, what causes the pressure on gauge 3 as the actuator cylinder is returned to the normal position?

6. In the meter-out circuit, what causes the pressure on gauge 3 as the actuator cylinder is returned to the normal position?

7. Describe the reaction of the actuator cylinder in the meter-in circuit when the pressure setting of the loading device regulator is varied (step 10).

8. Describe the reaction of the actuator cylinder in the meter-out circuit when the pressure setting of the loading device regulator is varied (step 23).

9. Do the circuits used in this activity provide any speed control during cylinder retraction? If yes, describe how. If not, illustrate how the circuits could be modified to provide control.

10. In your estimation, which of these control circuits provides the most accurate speed control? Explain why you selected this circuit.

Activity 18-3

Use of Special-Purpose Valves in Pneumatic Circuits

Name ______________________________ Date ______________

A number of special-purpose devices are used in pneumatic circuits to provide additional control and increase operating efficiency of a circuit. These devices range from a simple timing-volume reservoir that produces a time delay when using pilot-operated directional control valves to shuttle valves that automatically select the higher of two air pressure sources. This activity illustrates the application of one of the commonly used special-purpose valves: the quick-exhaust valve. It is designed to increase exhaust rates from a cylinder to allow the maximum extension or retraction speed.

Activity Specifications

Construct a simple pneumatic circuit designed to illustrate the effectiveness of a quick-exhaust valve in minimizing the retraction time of a circuit cylinder. Use the test circuits and procedures shown below to set up and operate the circuits. Collect the data specified in the Data and Observation Records section of the activity. Analyze the circuits and collected data before answering the activity questions.

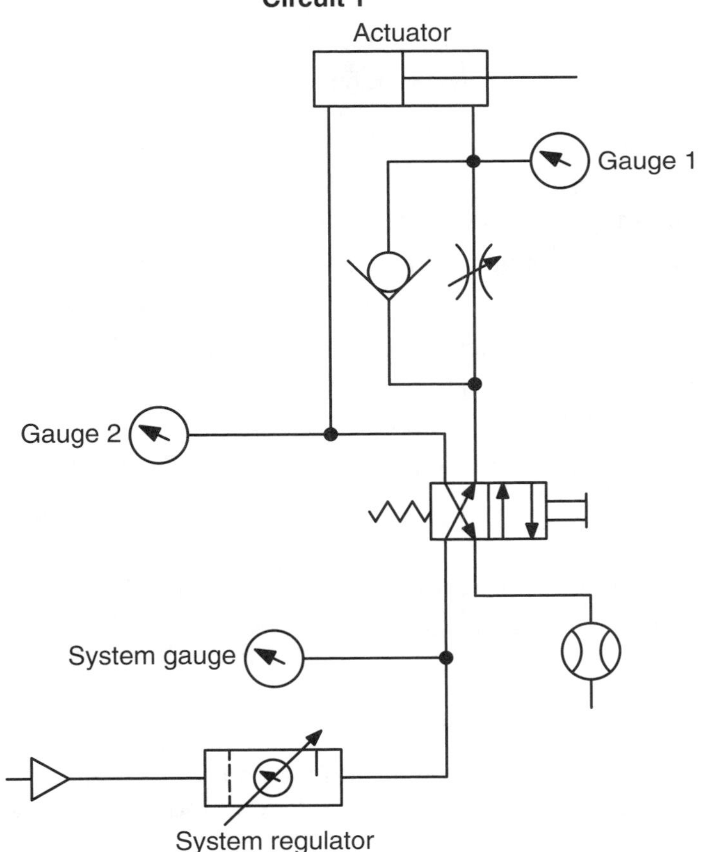

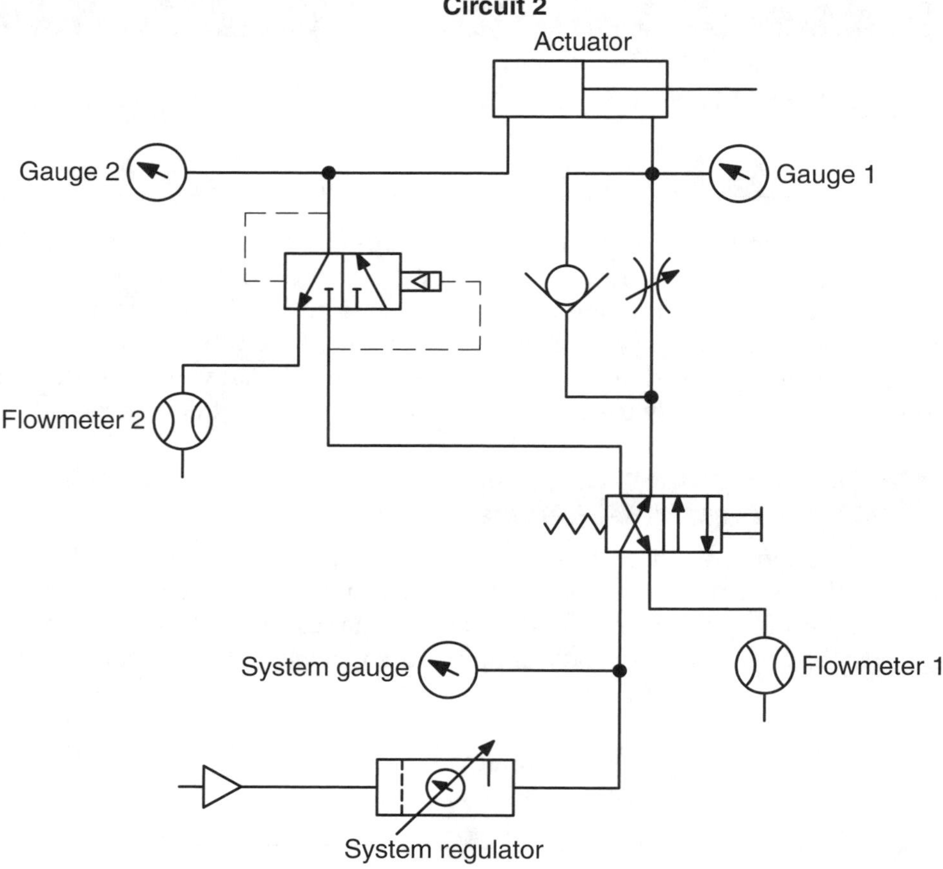

Required Components

Access to the following components is needed to complete this activity.

- Small pneumatic compressor unit or pneumatic distribution system workstation air supply.
- FRL unit (supplies high-pressure, regulated, conditioned air to the test circuit).
- Large-diameter, double-acting cylinder (used as a circuit actuator).
- Manually operated, four-port, two-position, four-way directional control valve (used to control the direction of movement of the circuit actuator).
- Two flowmeters (used to measure airflow exhausted from the cylinder).
- Quick-exhaust valve.
- Three pressure gauges (used to measure working line pressures).
- Manifolds/connectors (adequate number to make component connections).
- Hoses (adequate number to allow assembly of the test system).

Procedures

Study the following procedures to become familiar with the steps needed to complete the activity. Then, complete the procedures and record the observations indicated in the Data and Observation Records section.

1. Select the components needed to assemble circuit 1.
2. Assemble the system.

> **Note:** Be certain the flowmeter selected for the activity is of sufficient size to be easily read and to function without producing excessive pressure drop in the circuit lines.

3. Have your instructor check your selection of components and the setup of the circuit before proceeding.

Circuit operation without a quick exhaust valve

4. Start the compressor unit or turn on the main air supply.
5. With the directional control valve in the normal position, adjust the FRL unit to 30 psi, as read on the system gauge.
6. Shift the directional control valve to extend and retract the cylinder. Adjust the flow control valve so that the cylinder extension time is approximately 5 seconds.
7. Extend the cylinder. While the cylinder is moving, note and record in the chart in the Data and Observation Records section the flowmeter reading and pressures on the two gauges. Also, note and record the cylinder extension time.
8. Retract the cylinder. While the cylinder is moving, note and record the flowmeter reading and the pressures on the two gauges. Also, note and record the cylinder retraction time.
9. With the directional control valve in the normal position, adjust the FRL unit to 40 psi, as read on the system gauge.
10. Shift the directional control valve to extend and retract the cylinder. Readjust the flow control valve so that the cylinder extension time is maintained at approximately 5 seconds.
11. Repeat steps 7 and 8.
12. Repeat steps 9 through 11 to test the performance of the circuit using system pressures of 50 psi and 60 psi.
13. Check the chart and activity questions to be certain you have collected the information needed for this portion of the activity.
14. Disconnect the circuit from the compressed air supply.

Circuit operation with a quick-exhaust valve

15. Modify the test circuit to incorporate a quick-exhaust valve and second flowmeter, as shown in circuit 2.
16. Have your instructor check the connection of the quick-exhaust valve before proceeding.
17. Reconnect the circuit to the compressed air supply.
18. Repeat steps 5 through 12 to test the performance of the circuit using the quick-exhaust valve.
19. Check the chart and activity questions to be certain you have collected the information needed to answer all questions related to the two test circuits. Repeat any sections of the activity to clarify the operation of the circuits and quick-exhaust valve.
20. Discuss your general observations with your instructor.
21. Disassemble the circuit, wipe the components, and return them to their proper storage location.

Data and Observation Records

Circuit without quick-exhaust valve (circuit 1)

System Pressure Gauge Setting	Cylinder Movement		Flowmeter #1	Gauge Pressure	
	Direction	Elapsed Time		#1	#2
30	Extension				
	Retraction				
40	Extension				
	Retraction				
50	Extension				
	Retraction				
60	Extension				
	Retraction				

Circuit with quick-exhaust valve (circuit 2)

System Pressure Gauge Setting	Cylinder Movement		Flowmeter		Gauge Pressure	
	Direction	Elapsed Time	#1	#2	#1	#2
30	Extension					
	Retraction					
40	Extension					
	Retraction					
50	Extension					
	Retraction					
60	Extension					
	Retraction					

Activity Questions

Answer the following questions based on the cylinder retraction times, flowmeter and pressure gauge readings, and your observations during the operation of the test circuits.

1. What basic quick-exhaust valve operating concept allows the actuator in circuit 2 to retract at maximum speed?

____

____

____

2. Does the quick-exhaust valve have any direct influence on the operation of the circuit during actuator extension? Use data from the activity to support your answer.

____

____

____

____

3. Examine the flowmeter data from the chart for the circuit *without* the quick-exhaust valve. What actuator phase has the highest flow rate? Why?

4. Examine the flowmeter data from the chart for the circuit *with* the quick-exhaust valve. What actuator phase has the highest flow rate? Why?

5. Compare gauge 2 pressures during the retraction phases of both circuits. What happened to these pressures as the system pressure increased? Why?

6. Examine the *extension* pressures recorded for gauges 1 and 2 for each of the system pressure settings for both circuits. Why are these pressures very similar?

7. Examine the symbol for the quick-exhaust valve shown in circuit 2. Why does the valve shift to and *stay* in the position that extends the actuator when the manual four-way directional control valve is shifted to extend the cylinder?

8. What influence does system operating pressure have on the operation of the two test circuits? Use data from the Data and Observation Records section to support your answer.

Activity 18-4

Selecting Pneumatic Circuit Control Valves Using Manufacturer Recommendations

Name ______________________________ **Date** ______________

Selecting the proper size of control valves for a pneumatic system can be a difficult task. However, it is a task that is critical to the effective operation of a system. Components that are too small will not be able to perform the expected task or will operate at a reduced speed. Selecting components that are larger than necessary produces higher initial costs and higher-than-necessary operating expenses. This activity introduces valve sizing procedures that are recommended by valve manufacturers for their valves.

Activity Specifications

Select properly sized control valves for a specific pneumatic system using selection processes suggested by a manufacturer. The valve manufacturer selection process and the system specifications will be suggested by your instructor. Analyze the system specifications to become familiar with system operation and expected performance. Complete the selection process following steps suggested by the manufacturer. Carefully review any required calculations and make your valve selection before answering the activity questions.

Selecting and Sizing a Control Valve

Manufacturers often provide computer programs, formulas, or direct technical support for use in selecting appropriate components for a pneumatic application. These services are extremely helpful because of the nature of airflow in a system and the numerous designs and components available for an application. The following are examples of websites that provide various levels of support, including formulas, calculators, computer programs, or engineering support. Review these sites or other similar sites recommended by your instructor before continuing with this activity.

www.indianafluidpower.com

www.numatics.com

www.parker.com (select Products>Sizing and Selection Tools>Product Selection)

To complete this activity, obtain from your instructor a schematic diagram of a system containing cylinders, directional control valves, flow control valves, and lines that must be sized and selected. Also, obtain specific system operating parameters from your instructor. The following list can be used as a guide for operating parameters.

Weight of load to extend: ______________________________

Required time to extend load: ______________________________

Dwell time extension: ______________________________

Weight of load to retract: ______________________________

Required time to retract load: ______________________________

Dwell time retraction: ______________________________

System dwell pressure (psi): ______________________________

Cylinder bore: ______________________________

Cylinder stroke: ______________________________

Cylinder rod diameter: ____________________

Flow control valve port size: ____________________

Directional control valve port size: ____________________

Line material: ____________________

Line diameter: ____________________

Line length: ____________________

Line 90° elbows (number required): ____________________

Other: ____________________

Other: ____________________

Other: ____________________

Selection procedure

Use suggested formulas, programs, and other procedures from websites or provided by your instructor to determine the model and size of components needed to obtain desired system operation. Do *not* request engineering assistance from any manufacturer or sales organization to complete this activity.

Activity Analysis

Answer the following questions in relation to the calculations completed and the information collected.

1. Attach worksheets showing the formulas you applied and your calculations. Include copies of reference tables from which information was obtained during both the calculation and selection processes.
2. Describe the types of information that you identified on the Internet concerning pneumatic system design. Which materials were the most useful as you worked through the design activity?

____

____

____

____

3. Which website provided the most complete information that could be applied to the design exercise in this activity? Why was this information especially helpful to your task?

____

____

____

____

4. In general, can the formulas and design information found on the various Internet sites be applied to the components of other manufacturers? Use examples to support your answer.

____

____

____

____

Chapter 19

Applying Pneumatic Power

Typical Circuits and Systems

The following activities are designed to show the operation of several of the basic pneumatic circuits discussed in the chapter. The activities require assembly, operation, and analysis of selected circuits. These circuits show how a combination of basic components can achieve pressure control, speed control, automatic reciprocation, or increased system operator safety. The activities apply the function of individual components to operating circuits. Emphasis is placed on demonstrating concepts key to understanding the operation of a full range of pneumatic system designs.

> **Caution:** Check with your instructor to be certain you are working with the correct equipment. Also, follow all safety procedures for your laboratory as you complete the activities.

Key Terms

The following terms are used in this chapter. As you read the text, record the meaning and importance of each. Additionally, you may use other sources, such as manufacturer literature, an encyclopedia, or the Internet, to obtain more information.

automatic reciprocation ____________________

automatic return ____________________

boosters ____________________

composite symbol ____________________

logic functions ____________________

memory circuit ____________________

multiple-pressure circuit ____________________

quick-exhaust valve ____________________

remote pressure adjustment ____________________

time-delay circuit ____________________

trio unit ____________________

two-hand safety circuit ____________________

Chapter 19 Quiz

Name ______________________ Date ____________ Score ________

Write the best answer to each of the following questions in the blanks provided.

1. Name the three functions of the air-preparation unit located at the workstation.

2. The ______ of the air delivered by the workstation air-preparation units determines the force and torque of the system actuators.

3. When the concept of differential force is used to extend a cylinder, what are two limiting factors?

4. Which two factors need to be considered when designing a speed-control circuit?

5. Describe the design and placement of four-port, four-way directional control valves in flow control circuits.

6. Two cylinder extension speeds can be obtained by using mechanically or solenoid-operated, normally open, ______ valves to direct airflow through adjustable flow control valves.

7. A(n) ______ is basically an air-powered pump designed to produce pressure higher than the pressure available from the distribution line.

8. Describe the function of a *memory circuit.*

9. List the three key control elements of a *time-delay circuit.*

10. Briefly describe the operation of a *two-hand safety circuit.*

Activity 19-1

Construct and Test a Circuit That Controls Pressure from a Remote Location

Name ____ **Date** ________________

Placing a system control as close as possible to the circuit actuator often increases system performance. This occurs because reducing the distance the air must flow reduces the total resistance encountered in the system. The larger the system, the more the savings. This activity shows how a pilot-operated regulator can provide accurate control while reducing the pressure drop inherent in long air distribution lines.

Activity Specifications

Construct, operate, and test a pneumatic circuit that involves control of a pilot-operated pressure regulator and a system actuator from a remote location. The circuit should be constructed using the components listed and the diagram shown below. Collect data as specified in the activity procedures section. Carefully analyze the circuit and the data collected and then answer the activity questions.

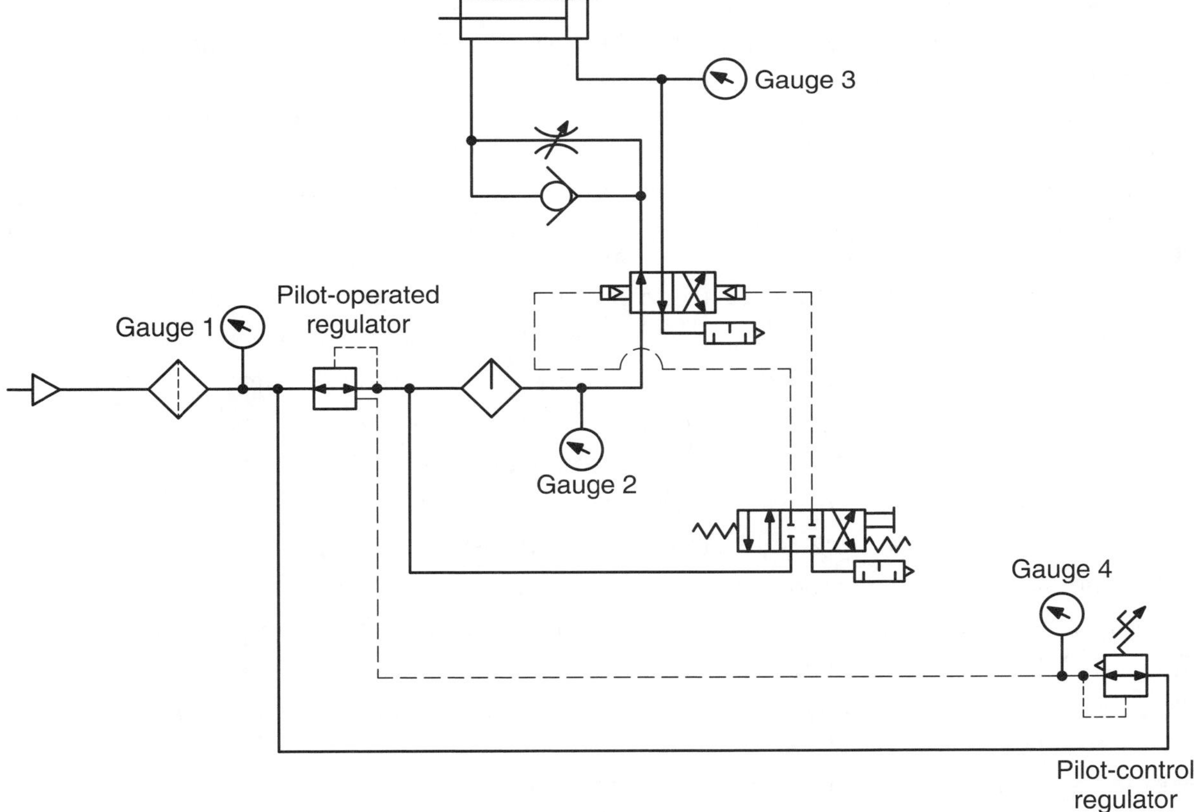

Required Components

Access to the following components as needed to complete this activity.

- Small pneumatic compressor unit or pneumatic distribution system workstation air supply.
- Filter (to filter air entering the test circuit).

- Lubricator (to lubricate air going to the cylinder and pilot-operated directional control valve).
- Large-capacity, pilot-operated regulator (to regulate the air pressure to the pilot-operated directional control valve and cylinder).
- Small-capacity regulator (to provide pilot pressure to the pilot-operated regulator).
- Double-acting cylinder (used as the actuator in the circuit).
- Adjustable flow control valve with a built-in check valve (used to adjust the extension speed of the cylinder).
- Pilot-operated, two-position, four-way directional control valve (used to control the movement of the double-acting cylinder).
- Manually operated, spring-centered, three-position, four-way directional control valve with blocked-center configuration (used to shift the pilot-operated, four-way directional control valve).
- Four pressure gauges (used to measure working line and pilot line pressures).
- Manifolds/connectors (adequate number to make component connections).
- Hoses (adequate number to allow assembly of the test system).

Procedures

Study the following procedures to become familiar with the steps needed to complete the activity. Then, complete the procedures and record the observations indicated in the Data and Observation Records section.

1. Select the components needed to assemble the circuit.
2. Assemble the system. Have your instructor check your selection of components and the setup of the circuit before proceeding.
3. Start the compressor unit or turn on the main air supply and adjust the pilot-control regulator until gauge 2 reads 60 psi.
4. Shift the manually operated directional control valve to extend the cylinder. Adjust the flow control valve to obtain an extension time of approximately 5 seconds. Several trial-and-error extension strokes will be required to make this adjustment.
5. Shift the manually operated directional control valve to fully retract the cylinder.
6. Note and record in the Data and Observation Records section the pressures on all gauges.
7. Shift the manually operated directional control valve to extend the cylinder.
8. While the cylinder is *extending,* note and record the pressures on all gauges.
9. At the end of cylinder extension, note and record the pressures on all gauges.
10. Shift the manually operated directional control valve to retract the cylinder.
12. While the cylinder is *retracting,* note and record the pressures on all gauges.
13. At the end of cylinder retraction, note and record the pressures on all gauges.
14. Be certain the manually operated directional control valve is in the center position.
15. Adjust the pilot-control regulator until gauge 2 reads 80 psi.
16. Check the extension time of the cylinder. Readjust the flow control valve to hold the cylinder extension time at approximately 5 seconds.
17. Shift the manually operated directional control valve to retract the cylinder.
18. Repeat steps 6 and 13.
19. Shift the manually operated directional control valve to fully retract the cylinder.
20. Quickly adjust the pilot-control regulator to several pressures other than 60 psi and 80 psi. Note the changes that occur on the pressure gauges and any air venting from the regulators.
21. Discuss your observations with your instructor.

22. Disassemble the circuit, wipe the components, and return them to their proper storage location.
23. Complete the activity questions.

Data and Observation Records

Pilot-Operated Regulator Setting	Manual Control Valve Position	Gauge Pressure			
		#1	#2	#3	#4
60 psi	Centered				
	Cylinder extending				
	Cylinder extended				
	Cylinder retracting				
	Cylinder retracted				
80 psi	Centered				
	Cylinder extending				
	Cylinder extended				
	Cylinder retracting				
	Cylinder retracted				

Activity Analysis

Answer the following questions based on the pressure readings and your observations during the operation of the circuit at the various pressure regulator settings.

1. What causes the pressure registered on gauge 2 during cylinder extension?

2. What does the pressure registered on gauge 4 indicate? How does this pressure relate to system operating pressure? Explain your answer.

3. What is the pilot pressure used to shift the four-way directional control valve that controls cylinder movement? Is this pressure level necessary to operate this valve? Explain your answer.

4. Describe two or more factors that justify the use of a pilot-operated regulator valve in a pneumatic system.

5. Why do you think separate filter, regulator, and lubricator components are used on this system rather than a combined trio unit?

6. What system pressure change is observed as the pilot-control regulator is quickly adjusted in step 20? Why do you think this happens?

7. Compare the physical size, capacity, and cost of the control components (pressure and directional) of this circuit to the same circuit applied to large industrial equipment.

8. In this circuit, can the lever on the manually operated directional control valve be released as soon as the cylinder is extending or retracting? Explain why or why not.

Activity 19-2

Construct and Test a Circuit That Provides a Time Delay for Actuator Movement

Name ______________________________ **Date** ______________

A means to time the movement of an actuator during system operation is often required in a pneumatic circuit. Actuator timing can be achieved using a number of different means, such as timing-volume reservoirs, electronic timers, or mechanical cams. In this activity, you will construct and test the effectiveness of one of these circuit designs.

Activity Specifications

Construct, operate, and test a pneumatic circuit that provides a specified time delay in the movement of a cylinder by means of a pneumatic timing reservoir. The circuit should be constructed using the components listed and the diagram shown below. Collect data as specified in the activity procedures section. Carefully analyze the circuit and the data collected and then answer the activity questions.

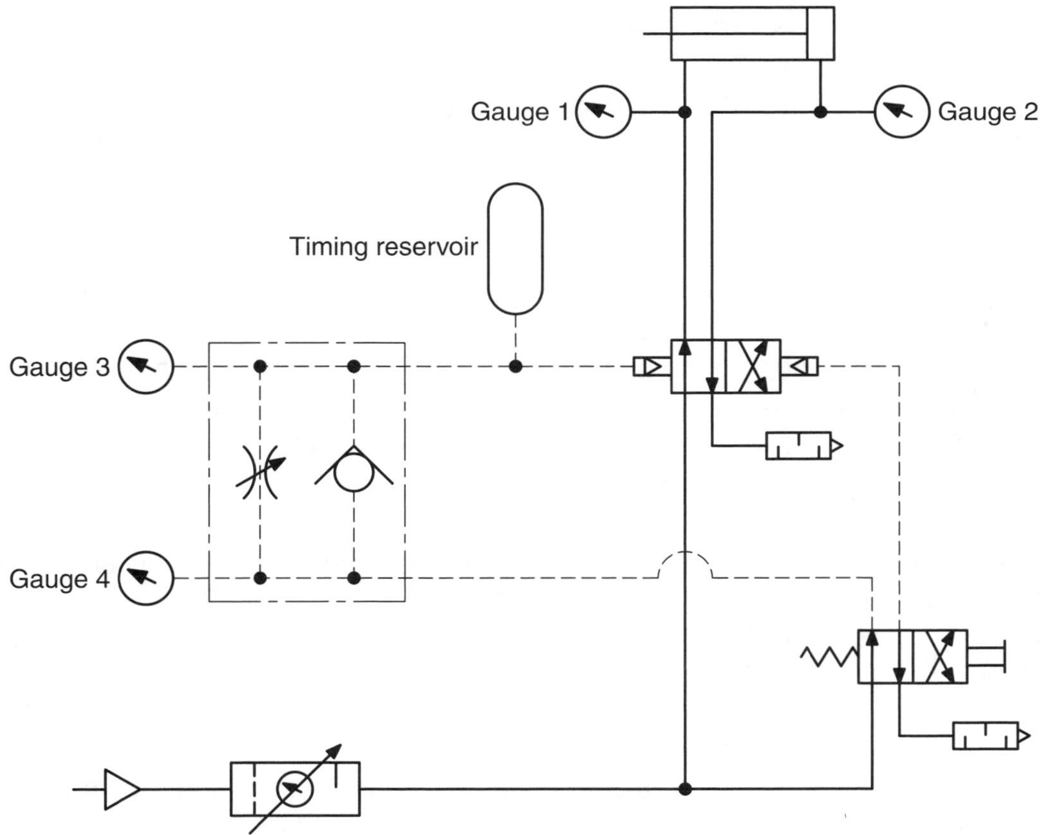

Required Components

Access to the following components is needed to complete this activity.

- Small pneumatic compressor unit or pneumatic distribution system workstation air supply.
- FRL unit (supplies high-pressure, regulated, conditioned air to the test circuit).
- Double-acting cylinder (used as the actuator in the circuit).

- Pilot-operated, two-position, four-way directional control valve (used to control the direction of movement of the cylinder).
- Manually operated, two-position, four-way directional control valve with spring offset (used to shift the pilot-operated, four-way directional control valve).
- Adjustable flow control valve with a built-in check valve (used to control the volume of air allowed to enter or exit the timing reservoir).
- Timing-volume reservoir (used to store pressurized air that provides the time delay for the pilot-operated, four-way directional control valve).
- Four pressure gauges (used to measure working line and pilot line pressures).
- Manifolds/connectors (adequate number to make component connections).
- Hoses (adequate number to allow assembly of the test system).

Procedures

Study the following procedures to become familiar with the steps needed to complete the activity. Then, complete the procedures and record the observations indicated in the Data and Observation Records section.

1. Select the components needed to assemble the circuit.
2. Assemble the system. Have your instructor check your selection of components and the setup of the circuit before proceeding.
3. Start the compressor unit or turn on the main air supply and adjust the system regulator to 40 psi.
4. Completely open the flow control valve.
5. Shift the lever-operated directional control valve to activate the circuit. Check the operation of the circuit to be certain the cylinder properly extends and returns to the retracted position when the directional control valve is released.

Set the automatic time delay between cylinder extension and retraction

6. Completely close the flow control valve leading to the timing reservoir. Extend the cylinder and then attempt to retract it.
7. Slightly open the flow control to allow air to bleed into the timing reservoir.
8. Adjust the flow control valve to obtain a time delay of 5 seconds before the cylinder retracts. Several trial-and-error cycles of extension and retraction will be required to make this adjustment.

Operate and test the time delay circuit

9. Allow the cylinder to return to the fully retracted position and the pressures in the pilot lines to normalize before completing the remaining steps.
10. Note and record in the Data and Observation Records section the pressures indicated on all gauges with the cylinder stopped in the fully retracted position.
11. Shift the manually operated directional control valve to extend the cylinder. Note and record any time delay before the cylinder begins to extend.
12. Note and record the pressures indicated on all gauges while the cylinder is *extending.*
13. Shift the manually operated directional control valve to retract the cylinder. Note and record the time delay before the cylinder begins to retract.
14. Note and record the pressures indicated on all gauges while the cylinder is in the *hold* portion of the cycle.
15. Note and record the pressures indicated on all gauges as the cylinder is *retracting.*
16. When the cylinder is fully retracted, observe the pressures in the circuit and compare them to those found in step 10.

17. Repeat steps 8 through 16 using a time delay of 15 seconds.
18. Adjust the regulator to 60 psi.
19. Repeat steps 8 through 17.
20. Experiment with circuit operation by reversing the cylinder before full extension. Note the general operation of the circuit under this condition.
21. Experiment with the circuit by using both shorter and longer time delays. Note any changes in the performance of the circuit.
22. Discuss your observations with your instructor.
23. Disassemble the circuit, wipe the components, and return them to their proper storage location.
24. Complete the activity questions.

Data and Observation Records

System Pressure	Delay Setting	Actual Delay	Cylinder Action/Movement	Gauge			
				#1	#2	#3	#4
40 psi	5 sec		Retracted				
		—	Extending				
			Holding (extended)				
		—	Retracting				
	15 sec		Retracted				
		—	Extending				
			Holding (extended)				
		—	Retracting				
60 psi	5 sec		Retracted				
		—	Extending				
			Holding (extended)				
		—	Retracting				
	15 sec		Retracted				
		—	Extending				
			Holding (extended)				
		—	Retracting				

Activity Analysis

Answer the following questions based on the pressure readings and your observations during the operation of the circuit at the various pressure regulator settings.

1. Briefly explain the basic concepts involved in the operation of this time delay circuit.

____

____

____

____

2. Does the cylinder extend without a time delay? Why or why not?

____

____

____

____

3. Draw a schematic of a modification that would allow a time delay on extension as well as retraction of the cylinder.

4. What is the function of the check valve in this circuit?

____

____

____

____

5. Why does the cylinder always return to the retracted position in this test circuit? Does this feature have anything to do with the operation of the time delay feature? Explain you answer.

____

____

____

____

6. What is the shortest and longest delay you can produce in step 21? How is the circuit/components modified to obtain these times?

____

____

____

____

7. What happens to the operation of the time delay feature when the cylinder direction is reversed before full extension?

____

____

____

____

8. Suggest at least two methods to extend the duration of a time delay or improve the accuracy of the timing. Be specific.

____

____

____

____

Activity 19-3

Construct and Test a Circuit Providing Automatic Cylinder Reciprocation

Name ______________________________ **Date** ______________

Processes in manufacturing and service equipment often require repetitive movement to complete a task. In many situations, this motion can be supplied by the reciprocating action of a double-acting pneumatic cylinder. This activity illustrates the design and operation of a circuit that provides continuous cylinder reciprocation.

Activity Specifications

Construct, operate, and test a pneumatic circuit that provides continuous cylinder reciprocation. The circuit should be constructed using the components listed and the diagram shown below. Collect data as specified in the activity procedures section. Carefully analyze the circuit and the data collected and then answer the activity questions.

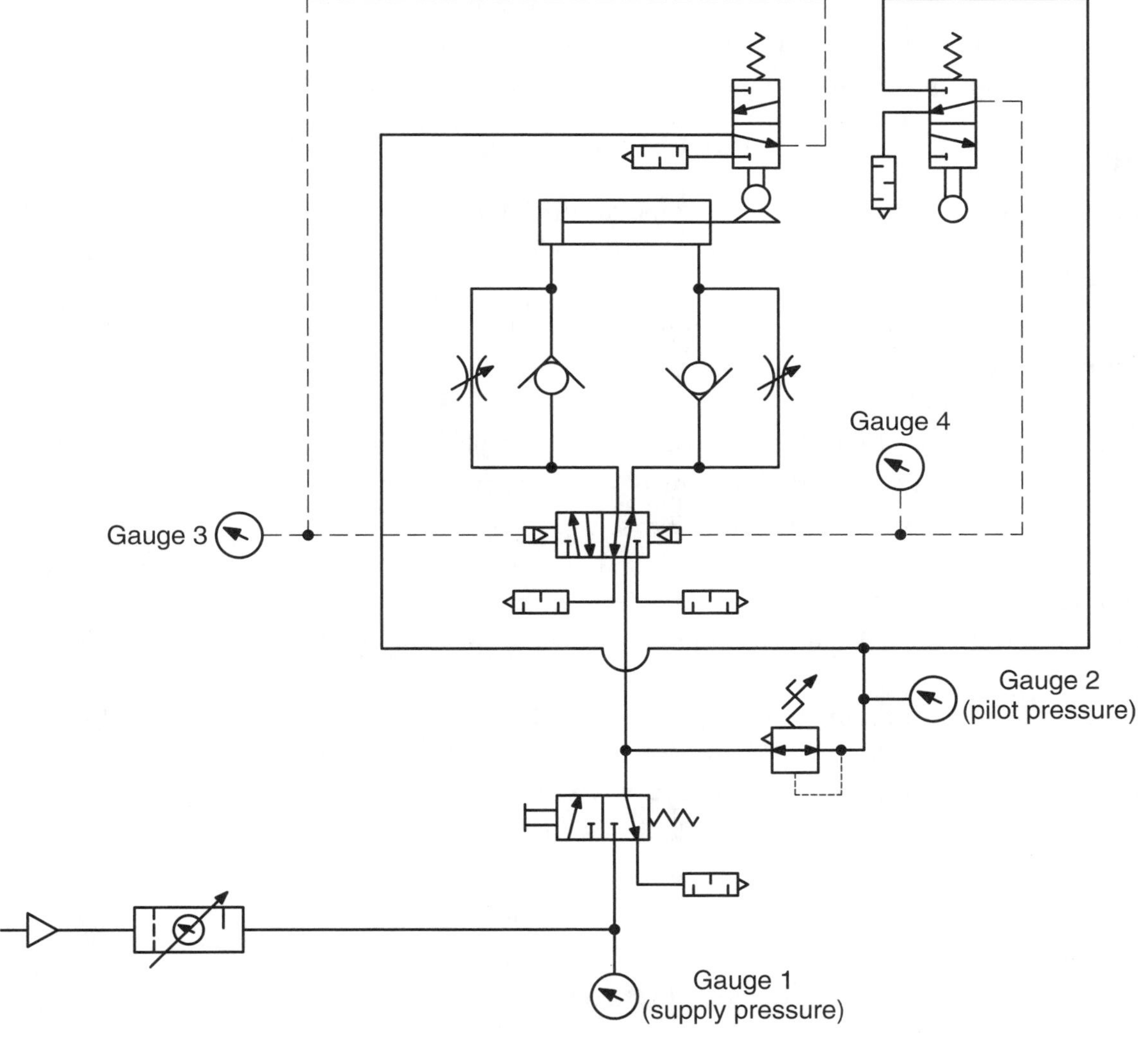

Required Components

Access to the following components is needed to complete this activity.

- Small pneumatic compressor unit or pneumatic distribution system workstation air supply.
- FRL unit (supplies high-pressure, regulated, conditioned air to the test circuit).
- Small-capacity, relieving-type pneumatic regulator (provides an additional pressure option to operate the pilot portion of the circuit).
- Double-acting cylinder (used as the actuator in the circuit).
- Cam ramp (mounted on end of cylinder rod to activate the cam-operated, three-way directional control valves).
- Two normally closed, cam-operated, two-position, three-way directional control valves (used to activate the pilot-operated, four-way directional control valve that controls cylinder movement).
- Mounting platform (device to hold the double-acting cylinder and two cam-operated valves in alignment).
- Two adjustable flow control valves with built-in check valves (used to control the extension and retraction speeds of the cylinder).
- Pilot-operated, five-port, two-position, four-way directional control valve (used to control the direction of movement of the cylinder).
- Normally closed, lever-operated, spring-return, three-way directional control valve (used to start and stop circuit operation).
- Four pressure gauges (used to measure working line and pilot line pressures).
- Manifolds/connectors (adequate number to make component connections).
- Hoses (adequate number to allow assembly of the test system).

Procedures

Study the following procedures to become familiar with the steps needed to complete the activity. Then, complete the procedures and record the observations indicated in the Data and Observation Records section.

1. Select the components needed to assemble the circuit.
2. Assemble the system. Have your instructor check your selection of components and the setup of the circuit before proceeding.
3. Start the compressor unit or turn on the main air supply and adjust the system regulator to 60 psi.
4. Shift the lever-operated directional control valve to activate the circuit.
5. Adjust the pilot-pressure regulator to 30 psi.
6. Adjust the flow control valves so cylinder extension time is approximately 6 seconds and retraction time is approximately 3 seconds.
7. Start and stop the cylinder several times by shifting the lever-operated, three-way directional control valve.
8. Note and record the system operating characteristics as the cylinder reciprocates. Use the chart in the Data and Observation Records section as a guide during your observations.
9. Decrease the operating pressure at the supply regulator to 40 psi.
10. Operate the system. Note and record the pressure readings as the cylinder operates.
11. Note any changes in cylinder operating speed.
12. Reduce the operating pressure at the supply regulator to 20 psi.
13. Repeat steps 10 and 11.
14. Reset the pressure at the supply regulator to 60 psi.

15. Shift the lever-operated directional control valve to activate the circuit.
16. While the cylinder is operating, adjust the pilot-pressure regulator to several pressure settings above and below 30 psi.
17. Note any changes in the performance of the circuit.
18. Discuss your observations with your instructor.
19. Disassemble the circuit, wipe the components, and return them to their proper storage location.
20. Complete the activity questions.

Data and Observation Records

Distribution Line Pressure	Cylinder Movement	Gauge Pressure				System Operating Characteristics
		#1	#2	#3	#4	
60 psi	Extending					
	Extended					
	Retracting					
	Retracted					
40 psi	Extending					
	Extended					
	Retracting					
	Retracted					
20 psi	Extending					
	Extended					
	Retracting					
	Retracted					

Activity Questions

Answer the following questions based on the cylinder retraction times, pressure gauge readings, and your observations during the operation of the test circuits.

1. What is the maximum pressure available to operate the cylinder at each of the settings of the supply pressure regulator? How closely does the pressure registered on gauge 1 match this pressure? Explain any variations that occur.

2. What is the duration of the pressure readings on gauges 3 and 4? What causes this type of operation?

3. Why does the pressure registered on gauge 2 during circuit operation remain relatively constant?

4. How does the performance of the circuit change as supply pressure settings are changed? Describe changes that occur as system pressure increases and decreases.

____

____

____

____

5. Describe the performance of the circuit as the pressure is varied during step 16. What would be the best pressure setting for the regulator controlling the pilot line pressure? Justify your answer.

____

____

____

____

6. This system can be operated using full system pressure for shifting the pilot-operated, four-way directional control valve. Identify at least one advantage and one disadvantage of using a separate regulator to control pilot pressure.

____

____

____

____

Activity 19-4

Construct and Test a Two-Hand Safety Circuit

Name ______________________________ Date ______________

The safety of machine operators is always a concern. Injuries occur when operators become distracted or take chances to increase the rate of production. A variety of devices and circuits are used with equipment to increase operator safety. This activity illustrates the design and operation of a pneumatic circuit that requires two hands to activate and continue the operation of a machine.

Activity Specifications

Construct, operate, and test a pneumatic circuit that increases safety by requiring the operator to place both hands on control valves to activate and continue operation of a machine. The circuit also contains a feature that prevents machine function if either of the hand-operated control valves is tied down. The circuit should be constructed using the components listed and the diagram shown below. Collect data as specified in the activity procedures section. Carefully analyze the circuit and the data collected and then answer the activity questions.

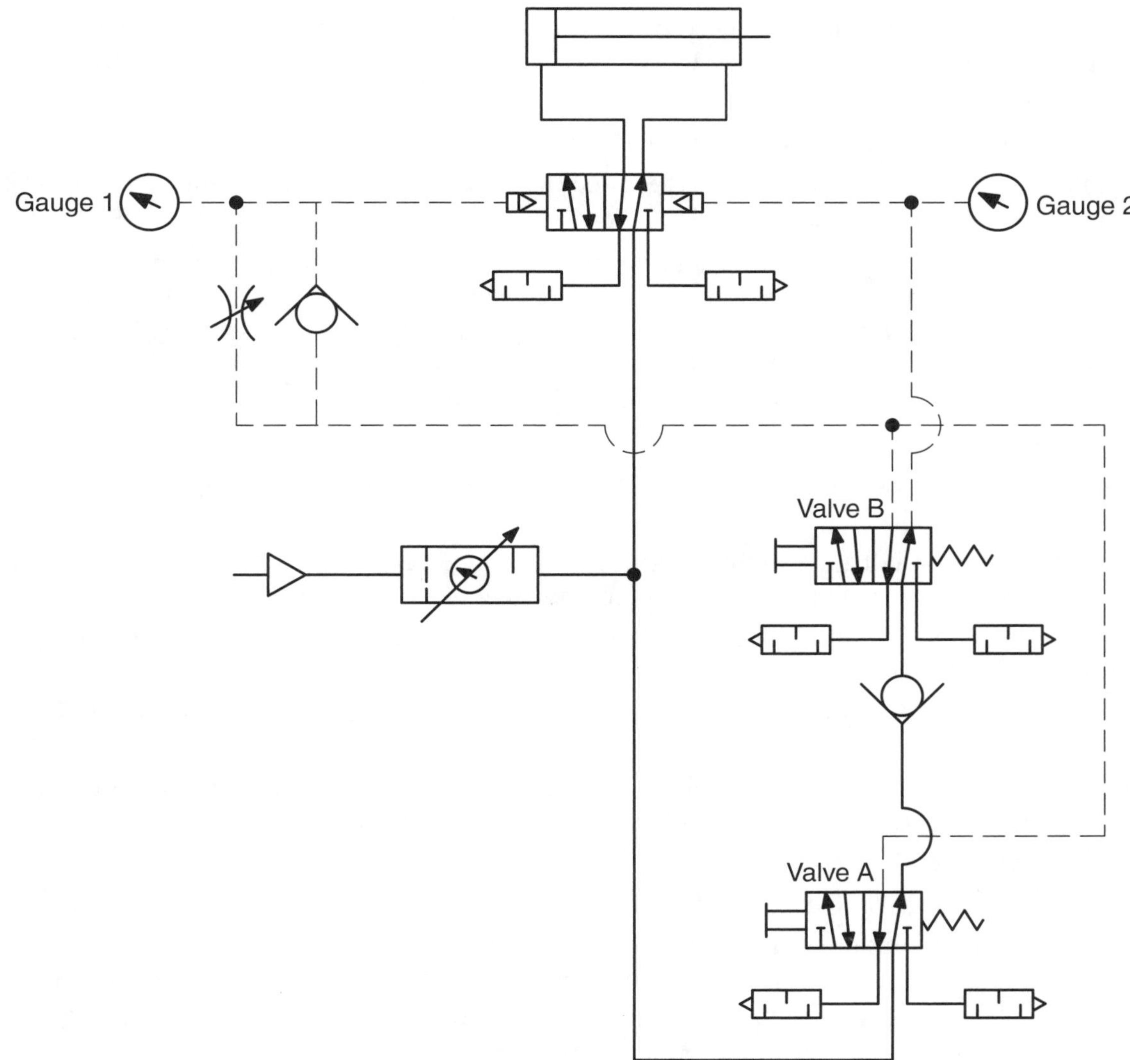

Required Components

Access to the following components is needed to complete this activity.

- Small pneumatic compressor unit or pneumatic distribution system workstation air supply.
- FRL unit (supplies high-pressure, regulated, conditioned air to the test circuit).
- Double-acting cylinder (used as the actuator in the circuit).
- Pilot-operated, five-port, two-position, four-way directional control valve (used to control direction of cylinder movement).
- Two manually operated, five-port, two-position, four-way directional control valves with spring offset (used to shift the pilot-operated, four-way directional control valve).
- Adjustable flow control valve with a built-in check valve (used to control the airflow into and out of one of the control chambers of the pilot-operated directional control valve).
- Check valve (used to prevent reverse airflow through the inlet line of one of the manually operated, four-way directional control valves).
- Two pressure gauges (used to measure working line and pilot line pressures).
- Manifolds/connectors (adequate number to make component connections).
- Hoses (adequate number to allow assembly of the test system).

Procedures

Study the following procedures to become familiar with the steps needed to complete the activity. Then, complete the procedures and record the observations indicated in the Data and Observation Records section.

1. Select the components needed to assemble the circuit.
2. Assemble the circuit. Have your instructor check your selection of components and the setup of the circuit before proceeding.

> **Note:** Be certain all lines are connected to the correct valve ports and the check valves are positioned in the proper direction. Special attention may be needed to tune the circuit for effective operation.

3. Start the compressor unit or turn on the main air supply and adjust the system regulator to 40 psi.
4. Note and record the position of the cylinder and the pressure on both gauges with neither valve depressed.
5. Depress the levers on both manually operated, four-way directional control valves to extend the cylinder. Note and record cylinder movement and the pressures on both gauges.
6. Release the levers on *both* manually operated valves. Note and record cylinder movement and the pressure on both gauges.
7. Depress the lever on directional control valve A. Note and record cylinder movement, the pressure on both gauges, and any vented airflow.
8. Release the lever on directional control valve A and depress the lever on valve B. Note and record cylinder movement, the pressure on both gauges, and any vented airflow.
9. Increase system operating pressure to 60 psi.
10. Repeat steps 4 through 8. Record the data in the Data and Observation Records section.
11. Tape down the lever on one of the control valves to simulate an operator attempting to bypass the two-hand safety system. Note what happens. Repeat the process for the other control valve.
12. Trace the flow of air through the pilot section of the circuit for each of the conditions identified in the Data and Observation Records section and the two taped-down situations.

13. Discuss your readings and observations with your instructor.
14. Disassemble the circuit, wipe the components, and return them to their proper storage location.
15. Complete the activity questions.

Data and Observation Records

System Pressure	Manual Valve Activation	Pressure		Vented Air		Cylinder Movement		
		Gauge 1	Gauge 2	Yes	No	Extending	Retracting	None
40	Neither depressed							
	Both depressed							
	Both released							
	Valve A only							
	Valve B only							
60	Neither depressed							
	Both depressed							
	Both released							
	Valve A only							
	Valve B only							

Activity Analysis

Answer the following questions about the two-hand safety circuit based on pressure readings, vented airflow, and your observations during the operation of the circuit.

1. Trace the pressurized airflow in the pilot control section of the circuit when both manual directional control valves are in the normal position at the start of circuit operation. Use a copy of the circuit diagram to trace the airflow.
2. Why must the check valve be placed in the circuit between the two manually operated, four-way directional control valves?

____

____

____

____

3. Why must the flow control valve be placed in the circuit between the pilot-operated valve and the manually operated, four-way directional control valves?

____

____

____

____

4. What happens when directional control valve A is taped down and an attempt is made to operate the circuit by shifting valve B? Explain your answer and trace the flow of pilot control air when this attempt is made. Use a copy of the circuit diagram to trace airflow.

____

____

____

____

5. What happens when directional control valve B is taped down and an attempt is made to operate the circuit by shifting valve A? Explain your answer and trace the flow of pilot control air when this attempt is made. Use a copy of the circuit diagram to trace airflow.

6. How quickly does the pilot-operated valve shift when the manually activated, four-way valves are released at the same time? Why?

7. Why is air vented from the pilot lines when only valve A is depressed?

8. Describe two situations where this type of circuit would be desirable on a piece of industrial equipment.